Anaconda

Anaconda

The Secret Life of the World's Largest Snake

JESÚS A. RIVAS

OXFORD

UNIVERSITY PRESS

OXFORD
UNIVERSITY PRESS

Oxford University Press is a department of the University of Oxford. It furthers
the University's objective of excellence in research, scholarship, and education
by publishing worldwide. Oxford is a registered trade mark of Oxford University
Press in the UK and certain other countries.

Published in the United States of America by Oxford University Press
198 Madison Avenue, New York, NY 10016, United States of America.

Library of Congress Cataloging-in-Publication Data
Names: Rivas, Jesús Antonio, 1964– author.
Title: Anaconda: The Secret Life of the World's Largest Snake/ by Jesús A. Rivas.
Description: New York : Oxford University Press, 2020. |
Includes bibliographical references and index.
Identifiers: LCCN 2019052497 (print) | LCCN 2019052498 (ebook) |
ISBN 9780199732876 (hardback) | ISBN 9780199753369 (epub)
Subjects: LCSH: Anaconda.
Classification: LCC QL666.O63 R58 2020 (print) | LCC QL666.O63 (ebook) |
DDC 597.96/7—dc23
LC record available at https://lccn.loc.gov/2019052497
LC ebook record available at https://lccn.loc.gov/2019052498

1 3 5 7 9 8 6 4 2

Printed by Integrated Books International, United States of America

I dedicate this book to my early mentors and friends. I will never be able to express how much I owe them:

Juhani Ojasti, El Veguero Finlandés;

John B. Thorbjarnarson, El Caimanero jodedor;

Stuart Strahl, who believed in me with no reason for doing so; and

Luis Levín; Quien hace una sinfonía con la ciencia

Contents

Foreword

Few animals in nature conjure up the mystique, awe, and fear triggered by mention of the anaconda, the most massive, and yet elusive, of serpents. Of the more than 3,500 identified snake species in the world, anacondas have received remarkably little study, especially in the field. Even in captivity, their size, dangerousness, and frequently pugnacious temperament have led to few scientific studies. We especially know little about their behavior, ecology, and conservation needs. In a world where humans seem intent on destroying biodiversity, altering climate, and modifying all aspect of nature that interfere with short-sighted human "needs" and growing populations, there remain, fortunately, dedicated scientists trying to understand the amazing life forms we still have on the planet. By doing so they aid in stemming the destructive tides of anthropocentric conceit. Those risk-taking pioneers studying the large animals most vulnerable to extinction are especially valuable in helping us appreciate the lives of charismatic animals before it is too late.

Jesús Rivas is a scientific adventurer and passionate student of reptiles, especially snakes and, in particular, anacondas. Gigantic anacondas (always females) have undoubtedly intrigued native peoples for millennia and westerners since the time of the conquistadors. Although frequently kept in zoos, their laid-back and largely inactive life there makes them seem dull, except for when they are on the attack, as in the laughably uninformed film *Anaconda*. Yet, outside of mating, described in detail in this book, perhaps the most dramatic behavior of anacondas is, in fact, capturing, killing, and swallowing prey. Yet most zoos do not show anacondas feeding on natural prey, let alone live prey of any type. So much for zoos being committed to real, not bowdlerized, education, although they do have to navigate complex public sensibility and ethical terrain.

This fascinating book gives us deep glimpses into how these animals live in nature. It combines detailed studies of the behavioral ecology and ethology of anacondas with the personal story of how Jesús developed as a scientist and field biologist along with his often provocative ethical and political views on research and conservation. However, the snakes, and Rivas's love and respect for them, are always apparent. Tasks necessary for his work such as locating, capturing, measuring, and sexing snakes; implanting transmitters; and navigating remote and often inhospitable environments were scientific keys to the project's success. Yet Jesús and colleagues also had to cope with injuries, come to terms with causing stress and pain to animals, and deal with the emotions of losing animals. Viewing and treating the anacondas as individuals with distinct personalities are no less appropriate for snakes than they were for Jane Goodall's pioneering approach to the Gombe chimpanzees. The detailed stories of Mamacita, Olivia, Lina, Marion, Penelope, and many others line up the scientific findings with the individual, much as how we expect medical doctors to view their patients.

Additional work beyond the individual is also necessary for successful field biology and natural history. How many animals live in an area, and where do they hang out as seasons change? What is the role of temperature, so important to ectothermic species? How does one find out what the animals are eating and how diet may change across size and season? What are the causes of death? How does the mating system influence the myriad aspects of their lives? What is behind the great differences in size between males and females? Anacondas have the largest female-biased size dimorphism of any terrestrial vertebrate. Is it only by chance they haunt the Amazon?

In addition to carrying out and supervising extensive fieldwork to gather basic quantitative and descriptive information, scientists need to systematize results, relate them to other studies, and, where possible, construct models. This is done in several areas of the book, for instance, in predicting the maximum size a female may reach, determining growth rates, understanding why some prey types are preferred over others based on size and nutrition, and comparing anacondas to other species. The section on the implications of the recent discovery of the extinct, and even more gigantic, *Titanoboa* in South America is riveting. Rivas also tackles issues rare in herpetological field studies and rarely discussed with any non-human species. What is it like to be the animal? Can we obtain some glimpse into their personal worlds, their Umwelt, as advocated by the Estonian biologist Jakob von Uexküll over 100 years ago but only now playing out in modern science? In short, do anacondas have a life worth living as sentient beings apart from any ecological role their existence serves?

In addition to these fascinating questions, research such as Jesús undertook entails numerous political, funding, and bureaucratic problems endemic to all long-term projects but more pronounced in remote fieldwork crossing national and cultural boundaries. As a Venezuelan, Rivas had a good grasp of governmental contacts, universities, and the Spanish language. Still, he had to deal with the urban–rural divide in the ranching culture of the llanos. His effectiveness in confronting and solving such challenges may have contributed to the success of the project as much as his scientific knowledge and empathy with the snakes. Indeed, dealing with people of numerous backgrounds and persuasions was necessary. Thus, there are numerous anecdotes of the many colleagues, supporters, students, and others who aided the work in Venezuela and labs around the world. Jesús is generous in acknowledging his many debts to numerous folks and institutions.

Part of Jesús's story is how he got to the United States and navigated the scientific and academic scene he found there, which often contrasted with the world he knew in Venezuela. This is told in the last chapter, along with briefly telling the story of how he became my Ph.D. student at the University of Tennessee. I had been working for many years in Panama and Venezuela on green iguana behavior and ecology. Indeed, some of our studies were in the llanos on ranches near those where Jesús worked. Surprisingly, while seeing much reptilian wildlife in the llanos, I had never come across an anaconda! This confirms that these gigantic animals are indeed not often visible in normal circumstances.

I was most fortunate to have worked with Jesús and immediately realized how creative, thoughtful, intense, and dedicated he was; soon his great potential as a scientist became apparent, along with his iconoclastic approach to phenomena that I have also pursued in my career. We bonded closely, although he was a voluble, outgoing Venezuelan and I derived from more taciturn northern Europeans. We both shared a lifelong passion for reptiles, especially snakes, from early childhood on. Jesús had also done important work on the behavior and ecology of green iguanas that I and my students, along with A. Stanley Rand and others, had carried out with the Smithsonian Tropical Research Institute in Panama as well as in the llanos of Venezuela. His discovery of a simple method to sex hatchling iguanas would have been invaluable in our work on hatchlings, where we did not have a nonintrusive sexing method.

Jesús points out that although his parents were educated professors, they were not happy when he gave up the prospects of a safe and respectable career as an engineer to study biology and especially the behavior of reptiles. "You won't be able to get any decent job." This resonated with me because decades earlier, in 1961, when I told my mother, excitedly, that although a chemistry major, my university had just established a biopsychology program that would allow me to study snakes and their behavior, the animals I had loved since early childhood, her first words, after a silence, were "How are you going to support a family?" I was crushed, and that conversation has never left me; and similar ones are indelible for Jesús as well. Ironically, this similarity in our lives I only learned about in this book! Fortunately, both families accepted our decisions; a memorable event in my career was meeting Jesús's mother when she came to Knoxville when Jesús received his Ph.D. and stayed at our cabin on Clinch Mountain.

The final chapters of the book provide a glimpse of the geological and climatic history of South America, especially the realm of anacondas. This helps us to understand their evolution and current distribution. This step back to look at the larger ecological and historical context of anacondas as they live today is informative. The narrative then shifts to current human pressures on anacondas and their habitats, political conflicts, and issues of wildlife management and conservation, including misplaced schemes that may do more harm than good as well as thoughtful suggestions for practices that may actually work. The criticism of economics and capitalism, especially in poorer countries, is severe. The stakes are great, and the goal is nothing less, as he ends this section, than a "peaceable economy."

This book is a rare treasure for anyone interested in the lives of large, potentially dangerous animals and the scientists who study them. A complement to this book is the recent one by Gay A. Bradshaw, *Carnivore Minds: Who These Fearsome Animals Really Are* (Yale University Press, 2017). She dives deeply and with great sensitivity into the world of other large, fear-inducing predators such as great white sharks, killer whales, grizzly bears, mountain lions, and crocodiles as well as some of the scientists who study them. With this book by Jesús Rivas, giant anacondas are added to the list of species with remarkable, if somewhat alien, minds and an equal right to persist on this planet as do apes, elephants, whales, bears, and sharks. Humans were responsible for the demise of many such animals in the past. These include the moa, dodo, auroch, elephant bird, great auk, cave bear, Steller's sea cow, passenger pigeon, Carolina parakeet, Tasmanian tiger, and giant monitor of Australia (which makes the Komodo dragon appear to be a dwarf).

The fate of large predatory animals in the world writ large is comparable to the tiny canary in the mine. If they succumb, are we far behind? We view our own species as mentally and morally vastly superior to other animals. If we do not heed the findings of those among us who provide tools and insights for solving problems facing the planet, our arrogance will continue to result in tragedy. Today many people despair that humanity, especially political leaders, will wake up in time, especially across a collection of disparate cultures stymied by conflict, ignorance, and misinformation but needing to work together. Nonetheless, our only hope is hope. In the meantime, the work of creative field biologists, brilliantly exhibited by Rivas, provides us access to the lives of remarkable animals and the tools to save them from ourselves.

I thus invite you to read this book, not superficially or quickly but taking time to absorb the many messages and the depths of data, logic, reasoning, empathy, interpretation, and wide-ranging information that you will encounter. Jesús points out that he wrote this book "with a thorough scientific knowledge of the anacondas resulting from years of rigorous scientific observation and inquiry" but also "with an unapologetic and very partisan point of view of somebody who loves the snakes and is honored to have had the chance to gaze into the secret world of the anacondas." Indeed, only through understanding the lives of others can we understand ourselves.

Gordon M. Burghardt

Acknowledgments

During several decades of studying anacondas I have received help and assistance from so many people in so many places that I'm bound to miss naming somebody whose help was crucial to my work and success. I apologize to them beforehand as I attempt to come up with a comprehensive list of my collaborators.

Perhaps I should start by thanking the five people who placed me on the path that I took. I will start out with John Thorbjarnarson, who gave me the opportunity to embark in this fascinating adventure that has become my life's work. We shared work, happiness, and misery doing fieldwork, and he was always a reliable, solid friend and mentor. I must also thank Stuart Strahl, who believed in me for no reason whatsoever, save for his great soul and trust in others. In his mind the only qualification I needed was to "enjoy the mud." I need to thank also Juhani Ojasti, who was a mentor and role model for most of my early career as a student in Venezuela. I am also in debt to Luis Levín, whose approach to teaching, research, and life was contagious and inspiring. He was always a source of moral support and strength even in the most trying moments of my career. I have no words to thank Gordon Burghardt, who accepted me as a graduate student when all metrics indicated that I was not going to be successful. In Gordon I found far more than a great advisor and mentor who let me err just enough to learn and guided my path and my career. I found in him also a great friend whom I admired constantly for his human quality and approach to life. When I came to work with him, I knew we shared a love of snakes and lizards, but I did not know we would share also beliefs in humankind, society, and even politics. My gratitude for these mentors can only be expressed with an aurora borealis.

I also thank the other member of my committee, Gary McCracken, Susan Riechert, and Sandy Echternacht, and other professors who helped me during my career, such as Gerardo Cordero, Jesús Ramos, Pedro Roa, Jose Luis Berroteran, and Frank McCormick, among many others who made important contributions in shaping my career. I thank Jesús Carrasquero for giving me the unbelievable opportunity to learn about snakes at Parque del Este. I am also very thankful for all the years at the Universidad Central de

Venezuela fire department where I acquired skill and experience far beyond what I would have elsewhere.

I also thank María Muñoz, who embarked with me at the beginning of this adventure. Not only was she a key player in the fieldwork and data collection but she also provided critical help and solid support dealing with all the non-science and petty political issues I had to face. I could not imagine what the first years of the project would have been like without her help, assistance, and personal support. Other people critical to get the project going were Bill Holmstrom, who knew more than anybody about captive anacondas; Paul Calle, the veterinarian who taught me to put transmitters in snakes and came to help with the fieldwork; Pedro Azuaje, who provided constant lo-gistical help in any and every situation; Saul Gutierrez, who recommended that I work in El Cedral; Mauricio Urcera and Alejandro Arranz, who were always there with help and support at Hato El Frío; and Mirna Quero, my boss at Profauna, who dealt with all the administrative issues. I thank Ramon Arbujas and Victor Delgado (*el viejo*) for help and lessons on finding anacondas in the wild. I also thank Dr. Kerian, the obstetrician/gynecolo-gist who kindly put his equipment and expertise to my service to study anacondas.

I am in debt for the abundant photographic material I received from photographers of the quality of Robert Caputo, Carol and Richard Foster, Tony Crocetta, Ed George, Philippe Bourseiller, Candido Luciani, Cristian Dimitrius, Roland Kays, Tony Rattin, Bill Holmstrom, Sarah Corey-Rivas, Paul Calle, Paul Rafaelle, Nick Lormand, María Muñoz, Ed Metzger III, and many others who selflessly contributed in more or less measure to the rich photographs in this book. I also thanks NMHU Media Arts department, and in particular Jesse Evans, for help digitizing photographic material. I am also very grateful to Luis Peralta, Brian Miller, Paul Andreadis, and Peter Buchanan for their feedback and insight on the writing of this book.

I also thank Profauna, CITES, the Wildlife Conservation Society for fi-nancial support, and Hato El Cedral (under the many owners) and Hato Santa Luisa, Hato El Frío, Agropecuaria Puerto Miranda, and the Doña Ana Society for logistical support of various kinds. Over the years I benefited from the help of a small army of volunteers, contributors, and students to whom I am very thankful. Among them are John and Theresa Dunbar, Roland Kays, Kako, Chip Foster, Mitch Barkasky, Torbin Platt, Stieg Klein, Paul Cowell, Frank Indiviglio, Rafael Ascanio, Peter Taylor, Courtney Andersson, Gerry Salmon, Christina Castellanos, Rich Zerilli, Mark Masio, Tibisay Escalona,

Rose Peralta, Reneé Owens, Steven Salinas, David Holtzman, Carlos Chávez, Gabriela Jimenez, Justin Saiz, Jennie Guilez, Christine Strussman, Juliana Terra, Vitor Campos, Peter Strimple, Bob Henderson, Neil Ford, Lars Holmstrom, Richard Elliot, Leora Peltz, Francis and Roseanna Faber, Xiao-Qing Qiu, Lisa McBride, Adrian Carter, Ana Caudillo, Rafael Ascanio, Rodney, Viviana Salas, Simon and Katz Tresseder, Andrew Travolacci, Victor Musiu Delgado, and very specially Sarah Corey-Rivas. To all of them, and many others, I am immensely grateful for their help.

1

Science, Love, and Anaconda Research

El primer horrible serpentón que se nos pone a la vista, por hallarse con frecuencia en aquellos países, es el buío, a quien llaman los indios jiraras aviofá, *y otras naciones y los indios de Quito le llaman* madre del agua, *porque de ordinario vive en ella.*

Es disforme en el cuerpo, del tamaño de una viga de pino con corteza y todo; su largo suele llegar a ocho varas; su grueso es correspondiente a la longitud; su modo de andar poco más perceptible que el puntero de los minutos de la muestra de un reloj. Dudo mucho que cuando anda en tierra haga en todo el día media legua de jornada; en las lagunas y los ríos, donde de ordinario vive, no sé a que paso nada; sólo el verle da notable espanto; y aun da consuelo saber cuán de plomo son sus movimientos. Con todo, el que sabe el alcance largo del pestilente vaho de su boca pone en la fuga su mayor seguridad.

(The first horrible large serpent that we find, because it is common in those countries, is the buío, to which the Jiraras Indians and other nations called *aviofá*, and the Indians from Quito called *mother of the water*, because it ordinarily lives in it.

It is deformed in the body, the size of a *viga* of pine including bark and all, its length often reaches 8 yards; its girth is correspondent with the length, its way of walking is a bit more noticeable than the minute hand of a clock. I very much doubt that when it walks on land it will make in a whole day more than half a league of travel; in the lagoons and rivers where it lives, I do not know at what pace it swims; the very sight of it gives noticeable fright; and it is even comforting to know how sluggish its movements are. Even so, he who knows of the reach of the fetid whiff of its mouth places in flight his greatest hope for safety.)

Excerpt from *El Orinoco Ilustrado y Defendido*[1, p. 376]

(Translated by JAR)

This book presents my personal account of the life of the anacondas after a quarter of a century of painstaking study of their biology and after having made my best attempts to understand what it is like to be an anaconda. I have learned much from the biology of the snake by rigorous application of the scientific method and the process of scientific inquiry in combination with state-of-the-art "gut feeling learning" and followed up with critical testing of my hunches. However, I did not set out to write a technical book where anacondas are studied from the outside in an "objective manner." Rather, I tell my personal account of what I have learned from these reptiles I love, over the last two and a half decades.

Like many of my fellow herpetologists, I never decided to become a herpetologist; I was one at birth. It was only later that I learned about snakes, reptiles, and other herps, but my love for them has been in me as long as I can remember. Growing up as a herpetologist in Venezuela has one big plus and a few minuses. The plus is, of course, the large number and ubiquity of snakes: You keep running into them whether you want to or not. The minuses were the lack of guidance and few possibilities to learn from them that I had growing up, going from having no material in Spanish that I could read, all the way to having no mentors or role models in college. I made a couple of unsuccessful attempts to do my undergraduate thesis with snakes, but I did not have enough support or guidance and ended up settling for a thesis on green iguanas, which are wonderful beasts themselves! I fell in love with iguanas and studied them for 5 years until an unexpected possibility fell right in my lap.

Between 1988 and 1990, international authorities confiscated 2,138 anaconda skins in an airport in the Netherlands that had come from Venezuela. Among them were skins from green anacondas (*Eunectes murinus*), which live in Venezuela, and yellow anacondas (*E. notaeus*), which live only in the southern parts of South America,[2] thousands of miles from Venezuela. This suggested that there was an illegal trade of anacondas throughout the continent. The approach that conservation organizations take when there is an illegal trade in developing countries is not to stop it by enforcing the law, but to develop a legal trade, which may be monitored and regulated, to drive the black market out of business. The Convention on International Trade in Endangered Species of Wild Fauna and Flora (CITES) stepped in and provided a grant to study anacondas in order to explore the terms and conditions under which a management program could be implemented. That is the opportunity that fell in my lap, and it was one I could not turn down.

A Dream or a Nightmare?

There I had it, my lifetime dream—the chance to study the most fascinating snake in the world. However, "management program" is code for harvesting, hunting, and killing animals for the commercial profits of a few. My study was intended to produce the baseline information needed to start an anaconda-butchering program! The reason I wanted to study anacondas so badly was that I loved them. Did I really have to come up with the terms of a program that would set up the sacrifice of thousands of them for profit?

I knew that somebody would do the research, if not me. So I chose to be in the loop to advocate for the snake rather than somebody else who would be looking out for the business side and not so much for the snake's interests. I could hitch a ride with their funding, and while I was answering their questions, I could seek to answer better questions.

The Questions of Science

At this point, I was drawn into the conflict of how much my personal opinions should permeate my research. Could I still be an objective scientist if I let my feelings for the animal determine what kind of study I did? In the big picture of biological research, there are two kinds of biologists: hypothesis-driven biologists and organism-driven biologists. The former are interested in testing hypotheses they feel strongly about. They would use any organism that is convenient to test their hypothesis—a plant, a bird, or a fish—so long as it meets the conditions they need for their study. They do not call their study organisms "snake," "bird," or "frog"; they call them "model" or "system." They design "smart" questions that can be answered quickly with the resources they have at hand, controlling all variables, often in the lab, and they produce the most publications for your buck. Lately, biologists have been leaning tremendously in this direction since it is consistent with the business model that has taken over the world, including academia and the scientific community, involving easy results and quick turnaround. You can set up an experiment asking questions on the mating system of 50 fruit flies in one afternoon, while I needed 9 years, and a lot more resources, to find 50 breeding aggregations of anacondas and tackle similar questions.

On the other hand, we have organism-driven biologists. They are often called "-ologist" as in herpetologist, ornithologist, entomologist, and so on.

Often the organism-driven biologists disguise themselves as "hypothesis-driven" in order to get grants and tenure, but when you see one you will know it. They cannot hide their real motive to study "their" organisms; their eyes shine when they see them or even talk about them. They have an emotional connection—they love their study subject and channel this love by learning about them. They call their study organism not "model" or "system" but names like Frodo, Diega, or Sandy, and it is not uncommon that they will adopt some of them as their personal pets.

Now, organism-driven biologists also ask testable questions and do hard science. The main difference between hypothesis- and organism-oriented biologists is that the former seek an organism that fits their question. Organism-driven biologists, on the other hand, find the organism they love and let the organism indicate what has to be studied about it. So, if you love *Ceiba petandra*, a tall tree that grows in the tropical forest, you are bound to ask questions about the emergent strata of the rainforest, or transport of nutrients, or light competition and epiphytes. If you love frogs, you are bound to ask questions about communication or worldwide extinction trends. If you love orchids, you will ask questions about co-evolution and symbiotic interactions. Whatever organism you love, you will listen to what the organism says and what it is telling you needs to be studied the most about it.

CITES had given me a very precise question: how to best harvest anacondas for profit. Was it the question that anacondas wanted me to study? Was it what anacondas were good for? How can the questions I ask affect my research?

How to Recognize a Good Question: Wearing the Snake's Shoes

So, I began a comprehensive project studying anything I could about anacondas in the wild. For the past 28 years I have studied them using the method of scientific inquiry. However, because of my conflict about objectivity in science, I have also learned a few things about scientific research itself. Any scientific endeavor is bound to have some bias of which the scientist may not even be aware. More often than not, our own biases are built into our way of thinking, the data we collect, and even the questions we ask. Trying to ignore our own biases does not make our research any more scientific. In fact, it can make it faulty if we have a bias that we do not account for in our analysis.

For instance, as I begin my narrative, I can hear some people disagreeing with the title of this book since it mentions that anacondas are the world's largest snake. I can easily hear someone claiming that some pythons reach comparable, or superior, lengths to those recorded in anacondas. However, if I ask, "What is the largest terrestrial mammal?" anybody who knows anything about terrestrial mammals would answer without hesitation: the African elephant. Nobody would start fussing about how much taller giraffes are. In mammals, as well as in any other vertebrates, mass is the component that we use to describe size, but when it comes to snakes, length seems to have a disproportionately higher value. Why is length considered more important in snake size than it is in the size of other animals? The answer may be that snakes come across as "lengthy" animals *to people*. Thus, the exaggerated emphasis that we give to the component "length" in snake size is actually a result of our own human perception. It is anthropomorphic! If we were to use the same yardstick with snakes to assess size that we use with other animals, we would use mass. Mass is also tremendously relevant in relation to prey size, predators, physiology, thermal inertia, and just about every other ecological variable we study. Anacondas are hands down the heaviest snakes in the world, and I thus consider them the largest (Figure 1.1).

So much for the question of size. Now, notice that the issue of snake size is as straightforward a question as one can hope to have. Yet herpetologists, scientists, and hobbyists alike have for years wrestled with this issue without realizing the damage that our own bias may have in our perception of the animals. If this bias overemphasizing the importance of length in snake size can confuse such a simple question, what happens when we address much more difficult questions? I have labeled this bias "anthropomorphism by omission" since we omit to put ourselves in the shoes of our study animal and fail to realize that we have a different perceptive world, priorities, and brains to process all the information.[3]

The fact is that our science can only go where our data and questions take it. Returning to the question of snake size, who cares what the biggest snake is anyway? Seriously, the only reason we ask about snake size is because we are fascinated with megafauna and large predators. These are human concerns and human interests. If we had the senses of some butterflies that can smell thousands of scents that we cannot detect, we would not be asking which snake is larger; rather, we would ask which snake smells more *sterckmorik*

(*sterckmorik* being a quality of scents that those butterflies love and that humans have no concept of). In fact, if we were butterflies, we might not even notice that snakes exist at all! For butterflies, snakes might not be any more than a geographic accident in the ground, no different than a rock or a tree buttress. Thus, the very question of which snake is larger has an element of anthropomorphism in it. Not knowing that we are being anthropomorphic may be even more dangerous to our science than embracing, and disclosing, any bias or emotions we may harbor toward the subject we study. I have thus immersed myself in the world of the anacondas, attempting to understand their world in their terms, not asking only the questions that I consider interesting but also trying to decipher what questions are relevant for their life.

Realizing how biased science can be without scientists even being aware of it, I have changed my position about the way I conduct my research. I have violated some of the prime directives of the scientific world about "objectivity," about not getting emotionally involved with my study subjects. As an example, I will tell you the story of Olivia, whom I caught the first year of the study (1992). She measured 4.91 meters (16 feet) but weighed only 43 kilograms (95 lbs). She was very long but very skinny, so we named her after Olive Oyl, the *Popeye* character. I installed a transmitter on her and followed her for 2 years. Day in and day out, I found her in the wild, and every day I did, there was a new little piece of the puzzle that she taught me about her life, mobility, diet, or habitat choice. I knew her whereabouts so well I could find her pretty much at will, without using the telemetry equipment. I kept running into her opportunistically years after the transmitter died, and every time it was like re-encountering an old friend who had something new to tell me about her life. Four years later, I found Olivia mating in a large breeding aggregation with five males. She was beautiful, strong, and bulky, a far cry from the original skinny snake I had found at the beginning of my study. I found her the next year again emaciated and with her body covered with bite marks from capybaras (*Hydrochoerus hydrochaeris*), which may defensively attack a predacious anaconda.[4] It was an interesting discovery and represented the first data I collected of a trend on the hazards that large females face after every breeding event.

A few days later, I saw a large spectacled caiman (*Caiman crocodilus*) that was tangled up in what seemed like mortal combat with a big snake. I prepared myself to enjoy the show that nature had set up for me that afternoon, and when I came up close I could recognize distinctive wounds and markings that I knew from Olivia's skin. She is getting a meal! I thought.

Then the caiman moved away due to my presence and I saw that the snake was not wrapping around the caiman anymore and a long rope of a snake followed the caiman loosely as it tried to leave, holding the snake's head in his mouth (this size a caiman had to be a male). Then I realized with horror that there was no combat: The caiman had Olivia square by the head and she had given up fighting. During my time with Olivia, I saw her killing caimans this size and bigger, but now she was wounded and weakened by the prior breeding event and she was in no condition to fight. An "objective" scientist would have collected the data on the death of snake E76C by caiman predation, but I could not sit and watch the caiman kill my old friend. I rushed into the pond and scared the caiman, forcing him to release Olivia. I put her in a cage to nurse her back to health. It was a long shot, because the caiman had sunk one of his long fang-like teeth right on top of Olivia's skull. Olivia struggled for her life for about 2 months, to no avail. I was crushed. There was no more wondering where she was. The mystery of when she was going to turn up again to teach me something else was there no more. She was gone. As I type these lines, I am wondering if there was not more I could have done to save her. Perhaps another shot of antibiotics, maybe pushing some more nourishment down her throat to help her fight infections. I will never know.

Should I have let the caiman eat Olivia and collected accurate data on predation of anacondas by caiman? I did report what happened as Olivia effectively being preyed on. I can make up all kinds of excuses of why the life of an individual that I have many data for is worth saving in the interest of the long-term understanding of the snake's biology, but that was not the reason I tried to save her. The reason is that I had an emotional connection with an old friend and did what anyone would do for a friend. The objective scientist would tell you not to let your emotions influence your work. Call me Venezuelan if you want, but I believe that the greatest accomplishments of humankind have been driven by deep feelings, compassion, and love. It would be silly to cripple our scientific efforts by removing the best that our human nature has to offer from scientific endeavors. I thus write this book with a thorough scientific knowledge of the anacondas resulting from years of rigorous scientific observation and inquiry, but I write it with an unapologetic and partisan point of view of somebody who loves the snakes and is honored to have had the chance to gaze into the secret world of the anacondas. I hope I can convey to you in the following pages some of what I have experienced and bring you a little closer to the secret life of the anaconda.

2

The Anaconda Challenge

Learning How to Learn

In order to learn about fishes one has to swim with them every day.

Francisco Mago-Leccia

I started the project funded by the Convention on International Trade in Endangered Species of Wild Fauna and Flora (CITES) and working for Profauna (the Venezuelan fish and wildlife service at the time) and the Wildlife Conservation Society, which also gave some logistic support. I also had the collaboration of two colleagues, John Thorbjarnarson and María Muñoz (Figure 2.1). If I had to mention two crucial aspects associated with our success, I would say: a lucky guess and a smart decision. The lucky guess was to study the animals in the llanos. When most people think of anacondas, they think of the Amazon rainforests where anacondas thrive. While it may be wonderful habitat for anacondas, it is a terrible place for people in terms of logistics, transportation, and simply getting by. If we had followed the natural tendency of seeking anacondas in the depth of the forest where they abound, we would have spent a lot of time and effort without much success. Instead, we started the study in Los Llanos, Venezuela's natural floodplains.

Los Llanos

Los Llanos is a flatland that takes up about a third of both Venezuela and Colombia, extending over 450,000 km² (174,000 square miles; about two-thirds the size of Texas). It is composed of an extensive system of natural, seasonally flooded grasslands. Gallery (riparian) forests bordering the rivers and patches of dry forest adjacent to them interrupt the otherwise continuous plain. Los Llanos is located to the north and west of the Orinoco River and sits on the northern borders of the Amazon basin (Figure 2.2). Because of this, most of the wildlife of the Amazon can be found in the llanos, where it is easier to observe animals in the open vegetation of the savanna (see reference 5 for a more detailed description of the llanos) (Figure 2.3). We counted on the help of a few cattle ranchers who were eager to protect the wildlife. Soon, Hato El Cedral came to our attention not only for its healthy wildlife but also for its good internal roads (Figure 2.4).

Upon arrival to the llanos one gets a good reminder of what tropical weather is. The temperature is high throughout the year and the seasonality is manifested mostly in the abrupt differences in precipitation. The area receives an average of 1,575 mm of rainfall a year (62 inches; similar to New Orleans) with over 90% of the rain falling between July and October, with torrential rains, resembling the monsoon rains found in the US Southwest, that often cause rivers to overflow and flood most of the savanna. However, these rains are not monsoon-like in origin. Rather, they are the result of the movement of the Intertropical Convection Zone. As the sun moves to the southern part of the northern hemisphere in the summer, it heats up the air masses close to the surface. These air masses rise because of the heat, and as they rise they cool down. As they cool, their moisture condenses and produces torrential rains. As the sun moves to the southern part of the southern hemisphere in the northern winter, it brings the rains there, leaving the llanos in a state of drought. The period between January and April is acknowledged as a dry season, with prolonged droughts that cause bodies of water to shrink (Figure 2.5). The 2 months between each season are transitional.[5] It was this extreme seasonality that made all the difference in the success of my research. Anacondas, being aquatic, concentrate in the few bodies of water that hold water during the dry season. During this time, the chance of finding anacondas was much higher than anyone anticipated.

El Cedral

El Cedral is a 54,000-hectare (133,000-acre) cattle ranch located in Apure state, Muñoz District (7°30′ N and 69°18′ W). A series of human-made dikes have created more permanently flooded habitats (*módulos*) where the impact of the dry season is diminished (Figure 2.6). The gates of the dikes are closed at the end of the wet season to hold the water for pastures and cattle. Each *módulo* has an extension of approximately 7,000 hectares (17,300 acres) that is managed with a low cattle density (0.2 to 0.08 heads per acre). This low density puts little pressure on the habitat and lets wildlife thrive on their land.

The dikes provide good and reliable roads to move around the ranch even at the height of the wet season. The construction of dikes (and roads) produced another kind of habitat that anacondas use and was advantageous for the study: borrow pits. Borrow pits are large holes left over from where dirt was taken to build the uplifted road. Accordingly, borrow pits are found along the roads and allowed us to find a "gold mine" of anacondas next to the road.

A Smart Decision

In January 1992 our team moved to Los Llanos to start the natural history project with the goal of discovering the terms for harvesting anacondas and with the intention of learning as much as we could about them. The basic tools we were going to use in the project were radiotelemetry and mark and recapture. I had never made such an irresponsible commitment before, and I have not since. With my two collaborators, we had 17 years of experience studying reptiles in the bodies of water of the llanos, but between the three of us we had had only three encounters with anacondas. To offer to study anacondas by radiotagging 12 of them in the first year, plus capturing *and* recapturing enough to have demographic results for a harvesting program, was downright irresponsible. I had no reason to justify, or to believe, that we were going to be anywhere close to successful.

The problem was that nobody had studied anacondas in the wild before, so there was no baseline knowledge for us to start from. There were people who had been working with anacondas in zoos for years. We had the invaluable help of Bill Holmstrom from the Bronx Zoo, who had great experience caring for a few anacondas he had in his collection. His help, while priceless when handling the animals, was of little help when it came to field ecology, where nobody had studied them.

Of course, there were people who knew a lot about them, just not in a formal, scientific way. Cattle ranching is done in Venezuela the same way it was done 200 years ago: the cattle roam loose on huge savannas and the llaneros (cowboys, the real deal, not just hats and boots) go out on the endless savanna and round up the feral living cattle that are ready to be sold (Figure 2.7). The llaneros (no boots at all, they often ride barefooted), after a lifetime riding up and down the savanna, were thoroughly acquainted with the anacondas. Although they did not like them, most had great respect for them, if not fear. They had plenty of knowledge that proved priceless for us.

I started my research by asking all the old folks about anacondas: what they knew about them, where they live, where they spend the dry season, how to find them, what they eat, how they breed, and every other question that came to my mind. Llaneros are naturally distrustful of foreigners (and yes, I was as good as a foreigner in these remote lands), and it took some doing to get them to open up. At first, they couldn't understand why I wanted to find anacondas: They did all they could to avoid them and often told me their scariest stories and reasons I should not venture out in the land of the anacondas. When I finally earned their trust, I found a plethora of good information about the animals that I could put to good use for my research (Figure 2.8).

However, there were a few issues I had to look out for. Llaneros, just like many other rural folks, do not think much of city slickers and amuse themselves with the naiveté of newcomers. Luckily, I was aware of this aspect of their style, and I managed to call them out when they were trying to put one over on me, which gained me more respect in their eyes and increased cooperation. Another source of errors I had to look out for was that most of what they knew was based on uncritical observations of the wild and from tall tales they had heard, so I couldn't take what they were saying at face value. Along with useful facts, there were myths and misconceptions. I worked out their stories with trial, error, and intuition until I gleaned the information I needed to know: how to find the anacondas.

I was surprised, though, that even the most unbelievable myths, which could not have possibly been true, were always grounded in at least a little bit of truth. One of the stories I heard commonly was about a snake that had antlers like a deer. As the story went, after reaching a certain size the animals grow horns and become enchanted snakes. Eventually I had the opportunity to find horned snakes myself! Not really. But I did see something similar. On occasions, anacondas suffer head wounds, and they may scar with an elevated keloid that resembles the nubs deer get on their head when they are growing antlers. Llaneros are familiar with these nubs in deer, so seeing something similar on a snake head, they assumed that the snake was in the process of growing antlers!

Getting Up Close and Personal with Anacondas

I learned from the cowboys that, in the peak of the dry season, anacondas hide under the aquatic vegetation where the water is shallow, and by poking with poles, it was possible to find them (Figure 2.9). Systematically, my team and I searched all the bodies of water (swamps, borrow pits, lagoons, and rivers) under aquatic vegetation, poking with poles and feet into the drying mud. We also found some animals in areas that maintain a lower temperature, such as caves on the river banks, large cracks on the bottom of the dry swamps, and spots protected from the sun under bushes next to the bodies of water.[4]

At first, we went in groups of two to seven people, with the help of volunteers from zoos across the United States. We walked under the baking sun feeling under the vegetation for what we could detect (Figure 2.10). Once somebody had detected a suspicious lump, we had to confirm if it was indeed an anaconda; often, to our disappointment, it was just a submerged log lying under the vegetation. On occasion, the tactile examination revealed the presence of a hard, smooth surface; this would indicate the presence of a savanna side-neck turtle (*Podocnemis vogli*). This was no reason to get excited, but no reason to worry either.

On other occasions the tactile examination revealed a hard, scaly crest that would suggest the presence of a spectacled caiman (Figure 2.11). These occasions were both disappointing and scary. The lump of a big caiman under the vegetation feels very much like that of a big anaconda, raising our expectations. However, our disappointment never lasted long when we realized we were touching a 1.8- to 2.4-meter-long (6- to 8-foot-long) crocodilian that was not being restrained by anything at all. Fortunately, caimans were surprisingly oblivious to being stepped on. The most reactive animals shook violently, sometimes throwing the person on his or her back, with an adrenaline surge running through their system. However, these animals never made any attempt to bite us (in no less than 200 encounters). In a similar incident, a 4.3-meter-long (14-foot-long) Orinoco crocodile (*Crocodylus intermedius*) (Figure 2.12) behaved in a similarly oblivious way when I stepped on it and felt its crests and flanks with my hands under the vegetation. My only guess is that these crocodilians under the hyacinth may be in some sort of seasonal torpor or aestivation that significantly lowers their defensive and aggressive behavior.

On happier occasions, we felt smooth scales with a soft cushioning feeling in the background. That meant that we were in business. Identifying the scales tells you there is an anaconda right there, but you have no idea what part of the anaconda you are touching (Figure 2.13). By feeling carefully, we could figure out in what direction the head was, based on the direction of the scales (they run toward the tail). However, this was not very helpful for figuring out the actual location of the snake's head. Somebody had to pull out a loop of the snake's body, and as it moved and thrashed around we could see where the head was and try to seize it.

This was just the beginning of the struggle. Big anacondas have few predators and seldom react defensively to mild levels of handling. But once you seize the neck, the snake realizes that she has to struggle (I use the female pronoun because all large animals are female)—and struggle they do. We then subdued them through physical exhaustion (Figure 2.14). Although strong, anacondas have an anaerobic metabolism that quickly depletes its energy reserves. After a short struggle the anaconda became too tired to continue fighting. We were wiped out as well after the intense struggle, but our mammalian metabolism recovered quicker and we could take the snake in for data collection and return her without her having time to recover. I recall how on the first days I ended up having sore muscles in my back, arms, and legs for days after we caught a big female anaconda (Figure 2.15).

Taking the Bad with the Good

The fieldwork was exciting, and although sometimes we went for days with little luck, when we finally found a big female it was worth all the hard work. Looking for snakes involves spending endless hours in the water, and in the early years, having my feet wet all the time gave me a very unpleasant foot rot (like athlete's foot, but well past the ankles!). I will never forget the expression of amazement of the dermatologist when I showed him my feet. He was alarmed and told me, "You have to keep your feet dry for 6 months if you want it to heal." To do this, I would have to walk on my hands, and I couldn't do that. Not being able to keep my feet dry to help heal my condition, my second best choice was to kick off my shoes and do the search barefooted (Figure 2.16). Because my feet would dry out immediately after coming out of the water, this not only helped to heal the uncomfortable condition, but it also gave me added sensitivity in my feet, allowing early identification of the diverse herpetofauna of the swamp. Clearly, being able to recognize the skin of a caiman before it got disturbed lowered my odds of having an unfortunate encounter with it. Also, when walking barefooted, you learn to tread lightly and to pick up your feet quickly if something seems dangerous.

There were many things that inhabited the waters where we searched for anacondas, from caimans, crocodiles, and electric eels to stingrays and piranhas. However, with the advice from the llaneros and my own experience, I knew how to avoid most problems and remain safe. Piranhas need plentiful oxygen in the water, so they do not like the still waters with a lot of aquatic vegetation where anacondas occur; the only danger was while moving through water without vegetation, and I could do that swiftly. Furthermore, the danger of these animals is overrated. Most times you need to be very unlucky to get one bite, and the feeding frenzies you see in movies or documentaries are rare and take a long time to develop. The only time some of us got bitten by piranhas was when I had stabbed my foot with the spine of a bush growing by the water's edge. As I walked in the water I chummed a trail of blood that attracted piranhas. Jen Moore, who had just graduated from college, was following me, walked into my invisible trail of blood, and was bitten by piranhas that had been attracted to the blood trail (Figure 2.17).

Stingrays are supposed to have an extremely painful stinger, but by shuffling while walking through the water it was possible to avoid stepping on them. Electric eels do resemble an anaconda under the murky water, and although a shock can leave you giddy for the rest of the day (I should know!), it is not that dangerous for a young healthy adult. As for caimans, I have already explained that for some reason they are not as much as a threat as they could be. There was, however, one animal that I was always worried about encountering, and I knew I could not avoid it. It is a tiny bug, a true bug of the order Hemiptera (possibly a Naucoridae, but I never had one that was not squashed for identification), that lives in the swamp and delivers the most painful, sharp localized sting. I named them MF bugs. As sharp as the pain is, it leaves no visible mark on the skin. It is hard to believe that you can have so much pain without any visual trace of it. After 40 seconds, the pain disappears completely, and when it is gone, it is hard to believe that something so painful is so completely gone, so quickly!

Secrets of a Snake's Head

Before we can understand what is going on in an anaconda's head, we need to know a little more about how it is built. Most snakes have adaptations of the skull and jaws involving mobile hinges, and a whole arrangement of joints and muscles evolved for swallowing large prey. Snakes gave up the solid bones and powerful jaws of their ancestral lizards for the benefit of increasing gape size (Figure 2.18). It would not be an exaggeration to compare the snake jaw and head to a high-tech device that can accomplish a very sophisticated function: being mobile enough to swallow large prey. Like any other high-tech device, it is very specialized and does not work well for purposes other than the one for which it is intended (in other words, never try to hammer in a nail with your TV remote). Of course, trading skull sutures and tight joints for mobile ones comes at a price. The lack of solid points of support and levers that allow for the extremely mobile jaws of the snake results in reduced bite pressure. The posterior migration of jaw muscles, needed for the large gape, makes them less able to deliver force (David Cundall, Personal communication). Thus, for the size a large anaconda has, it does not put a lot of pressure into its bite. I for one would gladly take the bite of an 5.5-meter-long (18-foot-long) anaconda over the bite of a 75-centimeter-long (2.5-foot-long) tegu lizard (*Tupinambis teguixin*)—or a Jack Russell terrier for that matter!

Another characteristic of high-tech devices is that they tend to be fragile and break easily. The extra mobility of the snake jaws is obtained by giving up the solid skull sutures that the ancestral lizards had, rendering the snake's head more vulnerable to damage. Because of this, an anaconda head can be easily injured while swallowing prey. María Eugenia, one of the females in our study, got a stick through the roof of her mouth that ended up in her orbital cavity (behind the eye). This probably was a result of swallowing prey that she had accidentally wrapped along with a wooden stick. Getting sticks stuck in their mouths is not an uncommon problem for snakes and can be fatal.[6] María Eugenia lived for years after this wound, but her eye was not functional. It goes to show how vulnerable these animals may be despite being top predators.

So we have this massive predator that has nothing to fear from most other animals in the swamp, but it has an extreme Achilles' heel. The head is probably the only vulnerable spot on the muscle-armored body of an anaconda. Because of this, anacondas are quick to protect it when they feel themselves in danger by hiding their head within the loops of their muscular body (Figure 2.19).

This vulnerable head is the reason that, when one catches a large snake, her first reaction is to pull a loop over her head to protect it. At first, this loop gave us all kinds of problems, to the point that we labeled it "the evil loop." More often than not, when handling a big anaconda, a person has one hand holding the head and the other hand holding (and held) by other parts of the snake's body. When this loop comes up from the snake's upper body to protect the head, it pushes forward the hand that holds the snake's neck, making the holder lose his or her grip. Then the snake proceeds to wrap the hand of the holder with her body. Having both hands wrapped by the snake, one ends up "held" by a snake that is now loose (see Figure 2.19)!

Over time, I learned to use this behavior to my advantage. A defensive anaconda is so worried about protecting her head that she does not try to wrap around or to constrict the holder, which she could. Her defensive instinct (a Fixed Action Pattern) is so strong that the snake would use up all her energy to protect her head before she tries to do anything else. Learning this was most helpful; at first we needed at least three or four people to handle a large anaconda, and the struggle was extremely intense for all of us. After learning how worried the snake is about protecting her head, I figured out that focusing on that first meter (3 feet) of a snake's upper body was all that I needed to wrestle the snake to exhaustion (Figure 2.20). Instead of trying to subdue 5-plus meters (18 feet) of wriggling beast, I focused on the top meter. By doing this, I (178 centimeters, 85 kilograms [5-foot-10, 187 lbs]) could dominate, by myself, anacondas much stronger than me and as long as 5 meters (Figure 2.21). We were also able to handle the animals with less force. Thus, the original "wrestling" became a kind of a tai-chi dance, using the exact movements needed to prevent the snake from developing its power and using them to control her until she tires herself out.

Bluff or Bite

I want to differentiate two terms that are often confused in the popular language or on TV documentaries about animals, in particular snakes. A loose definition of aggression refers to "unprovoked" attacks or behaviors an animal may display toward another, often associated with food, territory, mating, or other needs the animal has. When a snake reacts violently, perhaps biting, to being handled, it is defensive behavior, not aggression. This is an extremely important difference, because in the snake's world a bite is not always a bite.

Throughout the study I found a very consistent trend in the defensiveness of the snakes. Smaller snakes are much more defensive and disposed to bite than larger ones. This is a general trend among snakes, and in anacondas it is probably due to the lack of enemies of large snakes; an overly active response produces no advantage. If a large anaconda sitting in the swamp is disturbed or stepped on by something passing by, a capybara, caiman, or heron, it probably did not realize the snake was there. If the snake is not foraging, she has no reason to react to the presence of the other animal. So when we grab a snake while it is minding its own business in the swamp, it takes a while before she realizes that she may be in trouble. On the other hand, small animals that have a lot of predators, mostly males, bite very readily.

Recaptured snakes, too, are very defensive. They fight and try to bite, as if they now know that they have an enemy they did not have before. The behavior of recaptured anacondas is also similar to captive snakes. Many of my helpers were keepers from several zoos across the United States, and they all were surprised at how "calm" wild anacondas were, compared to the feisty disposition of captive ones. I believe this is also because, unlike their wild counterparts, captive anacondas know they are not the ultimate master of their domain and that there are enemies to be reckoned with out there.

Some people find it surprising that anacondas could remember former captures from year to year, enough to become more defensive in a recapture. This is perhaps part of that same human bias underscoring the abilities of non-human animals. I can recall an even more dramatic example of this. Joan is a big anaconda that I caught the first year and continued capturing year after year. A photographer friend, Bob Caputo, wanted to take some funny shots for the "On Assignment" section of *National Geographic* magazine. I had caught Joan and tired her out for data collection, so I suggested he could lie down on the ground and I could drape the exhausted snake on his chest for him to hold and pretend that he was wrestling her to take pictures. Because she was exhausted, I knew she would not be able to hurt him and a little extra handling would not entail much stress for her. Bob liked the general concept, and although he was uneasy about how safe he would really be, he eventually went for it. He held Joan by the head and pretended to be on assignment taking pictures up close while other collaborators and I snapped a whole role of film from different angles of the two of them. It worked well. Joan did not put up any fight and just let Bob handle her, never trying to squeeze or hurt him (Figure 2.22).

What I found most interesting was that as I moved around the two of them taking pictures, Joan kept rotating and using the little bit of movement that she had to keep me in her visual field. During the whole time she did not let me out of her sight and even tried to launch herself at me a couple times. In other words, she did not seem to be too worried about the person who was holding her by the neck but rather was worried about me! It is possible that she was keeping an eye on the other person since she knew where Bob was, but there were other people around whom she did not seem to mind. Along with the fact that she did not make any attempt to bite or move against Bob, I believe she knew that I, the person who had caught her for the last 6 years, was the one she had to look out for. Obviously, we will never know what Joan was really thinking, but her behavior was consistent with the idea that she had learned over the last 6 years that I was the problem.

While we were catching anacondas, often an animal became defensive and "tried to bite" the handlers. Often the snake launched herself toward one of us in a slow-motion fashion that allowed us to recognize her intentions and get out of reach. The first impression is that anacondas are so heavy that they cannot launch a really fast attack. However, after seeing the scorching speed in which anacondas may strike a prey, there is no doubt in my mind that all those "attempts to bite" were nothing but slow-motion bluffs of the snake trying to show us its reach and to encourage us to leave it alone. It's not unlike a boxer who jabs at his rival, trying to keep him at a distance but not really trying to connect a solid punch. I am convinced that the snakes never really intended to bite us until the last moment, when we held their neck or otherwise left them with no other choice than to fight.

This makes perfect sense. Recall how vulnerable the snake's head is and how worried she is about having it injured. Also, they have curved teeth that are intended for holding. It would not be very smart to bite a dangerous enemy and let their delicate head get stuck by its hooked teeth into such an enemy, knowing that their bite cannot do much harm. It would make much more sense to deter the enemy by bluffing.

Some Tricks of the Trade

On occasion, we did not leave the snake any choice but to fight, and while subduing an individual one of us would be bitten. Snakes' teeth are not intended to cut or to grind. Anacondas' teeth are like needles that curve backwards and are meant exclusively to hold. Thus, the wound was not necessarily a bad one, and anacondas seldom apply much pressure when they bite defensively. I learned through painful experience that the best thing to do was to resist the natural reaction to remove the bitten body part and instead reach with the other hand and seize the snake's neck; she often lets go immediately in an attempt to bite the seizing hand. But on some occasions (often recaptures), the snake would not let go even after being held. In these moments the bitten person must *never* pull back. This would only produce a worse injury on the hand and could even hurt the snake because the teeth come out very readily, leaving them stuck in the person's flesh. The tooth alveoli could get infected, leading to gingivitis. In these cases, pushing further down into the snake's mouth will be more successful, since the snake often opens the mouth when this maneuver is applied. If the person takes the right actions, he or she would get off simply with the painful imprint of all those needle-like teeth but not a serious injury.

Sometimes when handling a large specimen, the handler inadvertently holds the snake's mouth closed shut by applying pressure on the snake's head and jaws. This maneuver must be avoided because the snake's teeth do not oppose each other, unlike ours, and when one applies pressure to shut the mouth, the teeth will cut the opposing gum, producing bleeding and possible gum infections. Pay attention during the next snake documentary you see. If you see a snake whose mouth is bloody inside, there is a good chance that the handler made this mistake and has hurt the snake unintentionally.

Snake Muzzle

Once we had the snake subdued and under control, we had to collect data on it. Another important part of our learning curve was the development of a snake muzzle. While holding the anaconda's jaws closed, a cotton sock of appropriate size was pulled over the snake's head. Once the snake's head was all the way in (with the nose where the toes belong), we secured the sock to the anaconda's head with several loops of plastic electrician's tape firmly, but not tightly, around the snake's neck (directly behind the quadrate bone, Figure 2.23). Making sure to keep the mouth closed, we secured a second length of tape over the sock and around the snake's snout to secure the jaws, making sure not to block the nostrils. This tape must be placed snug but not tight. At this point the anaconda was muzzled and blindfolded, was behaving calmly, and could be released to be measured and to collect any other data we needed. The muzzle technique was very handy to work with the animals, but a muzzled snake cannot be left unsupervised lest the sock slip and the tape block the nostrils, suffocating the animal. Wet socks must be avoided too, as they may impede the snake's breathing.[7]

What Is the Length of a Snake?

Working with anacondas often represented a challenge in the least expected tasks, even for something as simple as measuring the animals. Most
herpetologists measure their snakes simply by stretching them on a ruler. It
is a lot different when the animal is too strong to stretch. Trying to stretch
an anaconda only makes her feistier and makes our lives more difficult.
We measured the animals by stretching a string over their back (following
the dorsal medial line) and then measuring the string (see Figure 2.23). It
sounds easy enough, but it turns out that, if the snake is struggling, measuring the same animal several times may result in substantially different
measurements. Sometimes taking wider or tighter turns as one follows the
dorsal medial line can make a big difference. We ended up measuring each
animal three times and considered the average as the "length" of the snake.[8]

The variation that we found in measuring the same animal led to several
insights, in particular regarding the question of how long a snake can grow.
On one occasion a well-known TV presenter came to the ranch with a film
crew while I was not there. Workers from the ranch who often help me look
for snakes found a large anaconda that they measured to be 5.5 meters (18
feet) long. My friends held on to the snake since they knew I was returning
soon. When I arrived the next week, they presented me with a fat anaconda that I measured to be 4.3 meters (14 feet) long: The snake had shrunk
more than 20% in less than a week! This was a sobering finding. We both
were measuring the same snake using the same technique, and we both are
scientists with an interest in finding the true length of the animal—and yet
there was such a large discrepancy in the size we recorded. In view of how
imprecise measuring a snake can be, certainly from person to person without
truly standardizing the methods, we should take a fresh look at the records
of snakes from the literature where we do not know how they were measured, or how cooperative the animal was, or how interested the person was in
assessing the true length of the animal (see Chapter 6). It seems to me that a
1 or 1.2-meter (3- or 4-foot) difference is not a meaningful one when measuring a snake longer than, say, 7 meters (25 feet).

A Personal Conflict

Despite my excitement over the opportunity before me, I began the anaconda project with a feeling of trepidation. There was an inherent conflict

between the mandate I was given and my feelings about the animals. My study was expressly intended to determine the best way to manage the anaconda population of Los Llanos. "Management" is, of course, a euphemism for killing wildlife for commercial profit. At the philosophical core of wildlife management is the assumption that harvesting a population for profit is the best way to protect it. When a species gains commercial value, the idea goes, there is financial incentive to conserve it. Case in point, lots of large wild mammals in the world are at risk of extinction, but cattle are at no risk at all of going extinct! While, in theory, it seemed to make sense, I had many misgivings about it, not the least of which is the personal feeling that there is an intrinsic value to wildlife, independent of the commercial use humans can make of it. Anacondas enrich the ecosystem by providing top-down control of herbivores and increasing diversity. They also enrich our lives, regardless of how much money per square foot their skins can fetch in the leather market.

However, the major source of my concerns went beyond my unease about killing the animals for profit. Profauna, Venezuela's fish and wildlife service, was at the time administering a major management program for the spectacled caiman (*Caiman crocodilus*). The caiman was the international poster child for how to best use wildlife resources, and thus represented the paradigm that all management programs should follow. Because of my experience with Profauna, I knew the program from the inside. I knew how sausage was made, and I knew all too well the unspoken glitches and "issues" the program had. I knew it was not at all a model we should try to mimic but rather something we should try to avoid, or fix. Profauna officials looked the other way when presented with tremendous evidence of corruption, overestimations of the numbers of animals, and countless evidence that the tanners were gaming the system. Profauna depended on the taxes the caiman program yielded and consistently disregarded the advice of the scientific community. I myself, or anyone working in the field, could see the decline in caiman populations long before it was fully realized by the Profauna authorities, and I knew I did not want the anacondas to meet the same fate (see reference 2 for a discussion on this matter).

This, in itself, was an important part of the reason I felt I had to get involved in the study and management of anacondas. I knew it would be far too easy to organize a sham program that gave the appearance of sustainable management, or even a conservation program, but was in reality a smokescreen to conceal a lucrative business for a small number of profiteers, simultaneously ignoring both the needs of the local people and the conservation of the animals and the environment. Ultimately, I decided to pursue the project so that I could be front and center in any national discussions of how to manage anacondas. I could be a watchdog for the CITES mandate, and of, course, I would be in a position to learn the secrets of this amazing animal. As chance would have it, this project ended up being my Ph.D. dissertation. Although I was not in any graduate program at the time, later, when I enrolled at the University of Tennessee, I already had 2 years of data collection and knew more about anacondas in the wild than just about anybody else.

3

What Is a Good Real Estate
for an Anaconda?

La llanura es bella y terrible a la vez; en ella caben, holgadamente,
hermosa vida y muerte feroz. Ésta acecha por todas partes; pero allí
nadie le teme. El llano asusta; pero el miedo del Llano no enfría el
corazón; es caliente como el gran viento de su soleada inmensidad.
Como la fiebre de sus esteros.

(The llano is beautiful and terrible all at once; it can fit, easily, side by
side beautiful life and atrocious death. The latter lurks everywhere;
but there nobody fears it. The llano scares; but the fear of the llano
does not chill the heart. It is hot like the great wind of its sunbaked
immensity; hot like the fever of its swamps.)

Excerpt from *Doña Barbara* by Rómulo Gallegos[9, p. 193]

(Translated by JAR)

Learning from My Dog

Following the excitement of having being chosen to lead such a ground-
breaking project came the realization that I had to do it. It was easy, when
applying for the grant, to say that I was going to catch lots of anacondas and
even recapture enough to make demographic models. But the truth is that
I didn't really know much of what I had to do. No one did. You just don't
run into anacondas all that often even in an area that has a healthy population of them. I would know it. Despite all my fondness for them and all
the time I had spent in the area studying iguanas, caimans, and crocodiles in
swamps, rivers, and lakes, where anacondas are supposed to live, I had very
few encounters with them.

Luckily, I had two secret weapons: radiotelemetry and folk wisdom. For
the first one I could put transmitters on a few animals and follow them so

I could learn from them. The second weapon would require a little more art, as we will discuss in the next chapter. A telemetry transmitter emits a beep; with the right antenna and receiver combo, I could find out what direction the beep was coming from and follow that line up to the very place where the snake was. So, I had the opportunity to follow the animals every day and get acquainted with what the life of the anacondas is like. It was like having a good friend that I could count on finding every day and learning what he or she was up to. Unlike human friends, they would not be very chatty or share their feelings over one too many drinks, but I could see where they lived, what depth of water they preferred, and what vegetation they liked and try to get a good feeling for what their life was like.

I had an advantage compared to other telemetry studies, which is that anacondas are not very skittish. Other people studying smaller animals, or mammals, need to worry about staying away from them so as not to disturb them while trying to learn as much as they can of their biology. Often, they have to triangulate the position of the subject from several locations and eventually figure out on a map where the animal actually is. Instead, I could walk right up to the place where the snake was, make visual contact, and see what it was doing without having to worry about disturbing it or altering its behavior, so long as I did not physically touch it. And even then, on quite a few occasions, I unwittingly stepped onto the animal I was tracking with little or no consequences on its behavior.

Sometimes I felt like I was doing the same thing my dog did. There wasn't any being in the world who knew me better than she did. She knew exactly what I was about to do, sometimes even before I thought of doing it, it seemed. How did she learn that? Simply by watching, watching, and watching. Of course, this is probably a bit anthropomorphic; watching me is what I saw her doing, but she was also listening and smelling, which probably provided her with far more information than her eyes did. But the basic fact is that she was paying a lot of attention to what I did. By observing intently and paying a lot of attention to all details, we can get to know quite a bit about a subject. Could I act like my dog? Could I watch and observe the behavior of a few animals with enough detail to draw relevant conclusions on the behavior of the species?

That was the advantage that using transmitters offered me, the possibility to make every encounter with an anaconda count the most. Sure, they were difficult to find. There was little I could do about that, but once I found one, I could wire her and make that encounter worth the most since the wired snake was out there for me to find at will so I could just find her and do what I learned from my dog: pay close attention to what she was doing. Of course, I was not limited to watching. I could collect data on the temperature of the water, the air, and even the animal, since some transmitters had a temperature switch that would broadcast to my receiver the animal's internal temperature. I could also document in detail what vegetation the snake preferred and just as importantly which ones she was avoiding.

In those days I lived in the field, an advantage for learning about anacondas that I didn't fully appreciate until I became a responsible professor at a university. Living in the field, I didn't have to teach classes, mentor students, or attend faculty meetings. My whole job was to find and keep track of these animals all day long, day in and day out. My data collection was not limited to planned field trips where I had to keep a tight schedule of activities to maximize the benefits of the short time I was in the field. This is what happens in most field projects that are based on field trips, as opposed to living in the field. My life mission was to learn what it was like to be an anaconda. Going to bed thinking of the last observation I made in the day and thinking of the animals I was going to observe next morning kept my mind focused on what I was seeing and allowed me to get a better feeling for what the life of the animals was like. Living there not only allowed me to keep up with the planned data collection but also granted me a lot of "idle" time in the field where I still had my eyes open and I could learn a lot more than I was planning. It opened the door for serendipity, which always provides the best pieces of knowledge in just about any research project.

As happens often in science, things are easier to plan than to execute. Wiring a snake sounds easy enough, but the problem is how to do it. Snakes have no neck to speak of, so collars are out of the question. Glues and external attachments are bound to fail when the snake sheds the skin. Harnesses and other methods used in quadrupeds are a joke for the legless and slender body of snakes; living in the mud, they are guaranteed to lose just about everything that you put outside their bodies. The solution had to be internal. Snake biologists had already developed ingenious mechanisms for surgically implanting a transmitter in a snake. Because I was working with the largest snake in the world, it seemed that the challenge for me was easier since I could implant a larger transmitter, with a larger battery, which would last longer and would give me more information about each specimen than was normal for other snake biologists.

However, with anacondas I was facing a lot of unknowns. My first challenge was to find 12 anacondas, the number of radios I had, to implant with transmitters. Given that in 17 years we had had only three encounters, the idea to have to find 12 for wiring seemed an insurmountable challenge. We were to wire four males and eight females, but we needed to find them first. We chose to split the number unevenly between boys and girls for a couple of reasons. First, a female that is breeding may be substantially different in terms of behavior, thermoregulation, and other behavioral traits than a female that is not breeding. Second, because males are smaller, we were not sure we could find enough males thick enough to receive a transmitter. Clearly, anacondas are large, in terms of length and mass, but a transmitter still needs to be thin enough so as not to impede the animal's movement, functioning, and internal organs, and not to hinder passage of food through the animal's body. This was not a problem with a 4-meter-plus (13-foot-plus) female, but it was different with males. We had to find males at least 3 meters (10 feet) long to be able to do the transmitter-implantation surgery safely, and this proved to be a challenge.

Casting the Role

Two important collaborators arrived from the Bronx Zoo (currently Wildlife Conservation Society). Paul Calle, a veterinarian, came to the field bringing equipment to do the surgeries he could do, and he would teach me the procedure so I could install the others. Along with Paul came Bill Holmstrom, the manager of the herpetological collection at the zoo, who had been breeding and taking care of anacondas in captivity for 17 years. He had the most experience handling anacondas of anyone in the world. Although his background was not from the field, he was quite an asset to have on my team when the study began and taught me a great deal about how to deal with anacondas.

The two men came for a short visit to help me get started. By the time they arrived I had a few females held in enclosures and ready to be implanted, but we had not caught any males large enough for a radio. So, we set out to find males. It was March, which we knew from the locals was anaconda mating season, so we felt we had a good chance of finding males moving around looking for females.

We suspected we could find anacondas in the area of the *módulos* containing deep water, but it is a complete waste of time to look for anacondas in deep water since our mobility is hindered and we just cannot detect them easily. It was more productive to drive along the road that separates the deeper part of the *módulo* from the one below. There were underpasses that, during the wet season, allow water to run down the natural slope of the landscape and prevent it from rising above the level of the road. If it did, it would erode the dirt road away, making it impassable. It would also make it impossible for the upper *módulo* to hold water when the dry season resumes. However, in the dry season the underpasses are like small caves protected from the baking sun, and they often hold more moisture and are cooler than the rest of the savanna, creating a coveted refuge for many animals.

Shortly into our quest we came across a pretty large male, some 3 meters long, that was crawling right by the side of the road, heading down into an underpass. We stopped the vehicle and quickly had the male secured and under control. Because we now had one big male in which we could implant a transmitter, we celebrated loudly in the spot.

John believed that this male might have been following the scent of a breeding female enjoying the cooler and moister environment of the underpass. He was wrong. In the extreme heat of the dry season, a dry cave is a desirable spot for just about any animal who wants to escape the heat. As John approached the underpass looking for a breeding female anaconda, instead he found a nest of very angry, and very hot, Africanized bees! A buzzing cloud swarmed out of the underpass just like you might see in a cartoon, chasing everything that moved. We all scattered through the savanna. Paul, John, and María ran down the road as fast as they could. Bill and I happened to be closer to the other direction, so the obvious choice was to run up that way. Staying on the road was the best way to avoid tripping in a hole and falling. However, I was not about to let go of my prize: the one large male anaconda able to carry a transmitter that I had caught to date. So, I dashed up the road as well as anyone can run carrying a 3-meter-long anaconda. A large constrictor tangled up around your legs and trunk can really cramp your style when you're trying to break into a gallop! But the bees buzzing in my ears and a few stings gave me that extra incentive to get my legs moving. I just sucked it up and ran, carrying my snake, as best as I could.

Not 100 yards into our escape, we saw by the side of the road (the same side where we had caught the male) another good-sized anaconda that looked like a second large male. It was like Providence was playing a prank on us: For more than 2 months we had been unable to find any large males to implant, and then, while we were being chased by 40,000 or 50,000 enraged bees, we found two of them! I pointed it out to Bill, who stopped. He hesitated for a few seconds, trying to decide whether to continue running or pick up the second male. Despite the increased buzzing and stings, Bill decided to brave it: He ran to the side of the road and picked up the male, and we both resumed our flight, now holding a big male anaconda each.

It so happened that running up the road, we also ran against the wind. Lucky for us! The dry season's strong surface winds were blowing in our favor.[10] The bees need to outfly the winds in order to chase us. You can't run as quickly against the wind, but since we had solid contact with the ground, every step counted. It was harder for the bees to chase us than it was for us to escape them. A few hundred meters more, running as best we could carrying one anaconda each, and we were completely free of the bees.

We stopped to rest and celebrate our double triumph, finding not one but two of the prized male snakes and escaping the bees. We were happy we had gotten away with our daring escape and Bill's even more daring capture of the male under the cloud of bees. Then we realized, out on the horizon, that our colleagues were downwind and were still frantically running full speed and beating themselves on their heads. While the sting of a killer bee is no worse than that of a regular bee, a person who receives many stings could be in serious danger, even if he or she is not allergic. From a previous conversation I knew that John was allergic, so our celebration was abruptly cut short. The problem was that the vehicle we had to get back to camp, or to the hospital, was parked on top of the underpass containing the enraged bees.

I tried twice to reach the vehicle so I could pick up the rest of the crew, but approaching them in the direction the wind was blowing the bees detected my presence shortly after I tried to approach them and chased me away from their nest. Worried about our teammates, I gave my snake to Bill, who thankfully was strong enough and had enough experience to hang on to both snakes at the same time. My only choice to get to the truck was to go around the swamp and sneak up on the bees from downwind. By the time they saw me, I was close enough to make a dash for the truck. The keys were mercifully in the ignition, and the old truck, which often hesitated to start, that day started instantly! I floored it and soon was down the road meeting the rest of the team, who, thank God, were tired, stung, and shaken but OK. By the time we returned for Bill, he was about to lose the battle with the two snakes. The poor thing had a head on each hand and was wrapped up in legs, arms, and trunks. The snakes slowly but surely were getting the upper hand, but Bill managed to hang on to them until we arrived.

That was the first time I met Bill, but soon I learned that he had quite a talent for getting himself into trouble. One day looking for anacondas, the whole team was walking by a shallow pool of a river that was no longer flowing, in the dry season. Someone detected undulating movement right under the surface of the murky water that showed a long, cylindrical, snake-shaped animal swimming right under the surface. The head of the animal broke the surface, and someone saw the unmistakable orange bands that anacondas have behind the eyes. The turbulence of the water as the shy animal withdrew underwater did nothing to hide the fact that its head alone was at least 12 centimeters (5 inches) across—a good-sized snake! Everybody rushed to the scene to catch the snake, but Bill had gotten a head start and had reached the snake ahead of everyone else. He plunged his hand into the water where the head of the snake was last seen—only to grab the solid, slick, and electrified body of a 2-meter-long (7-foot-long) electric eel! Of course, Bill had to let go, startled: A 500-volt shock, of low amperage, is not lethal but can really wake you up in the morning. As Bill recovered from the surprise in the small pool of water, he stumbled to the side a few paces, only to step on the eel again and get a second shock!

Looking for and catching anacondas required hands. Before I learned the tricks of how to subdue a big animal, numbers were the only way to go about catching a big snake (Figure 3.1). So, I was always on the lookout for helpers. One day two adventure-seeking German youths stumbled upon my field site. They had come to Venezuela looking for a place to camp in the tropical wilderness. I told them about my work, and since they were so enthusiastic, I enrolled them in my team. They did not get to camp much, since I hosted them in my modest accommodation, but as far as adventure goes, they did get a bit of it.

Right behind the station ran a shallow creek that stopped flowing during the dry season but held water throughout the year. Intermittent parts dried out while other parts held pools that were up to 1.2 meters (4 feet) deep in some places but were mostly 40 to 60 centimeters (16 to 23.6 inches) deep. The creek, Caño Guaratarito, had a bit of a gallery forest in one part where there was a bank, but the lower part of the river flowed through a savanna without banks or trees. It was apparent to me from the beginning that the lower part of the river had fewer anacondas than the upper part. The explanation was not immediately obvious to me. It was the same river, with a similar amount of capybara, caiman, and deer in both places. The more forested part may have had other prey, but from what I was learning about the diet of anacondas, they were not important players.

One day, looking for snakes in Caño Guaratarito, I spotted something that looked like an anaconda skin breaking the surface in a patch of water hyacinth in the middle of the river, right in the transition between forested and not forested. I approached the area only to see with disappointment that the snake whose skin I had spotted was a dead animal, and a pretty big one too. Going closer to inspect the corpse, I noticed the round body of a pretty thick live snake that I almost missed as it was covered by mud. I reached to pull it out to discover it was the tail of a massive anaconda under the mud, next to the dead one (Figure 3.2a). Excited, I grabbed with both my hands but could barely get a grip on the snake, which was easily 30 centimeters (1 foot) across. As I tried to pull her, I saw something else steering under the mud. I held the big snake coil with my elbow against my chest to free one hand so I could reach for whatever was steering under the mud. I felt my heart pounding when I realized that it was the scaly body of another big snake. I grabbed it and pulled it out, calling for the rest of the team to come to my assistance. There was no way I could manage one of those beasts by myself, let alone both of them.

It took a while for the snakes to realize that there was a problem. They didn't move much at first, but once they realized something was up, they started trying to pull out of my grip. With the girth of their bodies preventing me from establishing a firm grip and the slippery mud covering their scaly skin, there wasn't much I could do to hold on to them. I let their bodies slip between my fingers up until the point that my fingers got around the body, toward the end of their tails. Then I tightened my grip there, hoping against hope that I could hang on to them. As I tried to stabilize myself to resist the combined pulling of the snakes, I realized that the ground underneath me was moving and was not very stable at all. Soon I realize that I was standing on a carpet of slithering anacondas. I tried to pin down as many as I could between my knees and wrist without losing the ones I already had. I put the tail of two of the smaller ones behind my knee and by crouching was able to hold them between my thigh and my leg. My hands were holding on to the big tails, and my arms were pinning more animals against my trunk. With my weight I knew I was holding some more snakes, but I had no idea how many there were or whether I was holding one snake twice!

María was the first one to arrive to help, but even our combined strength and arms were no match for all the snakes we were trying to capture (Figure 3.2b). We were about to completely lose the battle when Bill and the German helpers arrived. Bill and I focused on holding the snakes in place, and the Germans worked on bringing them one by one to the shore, where María did what she could to secure them. Bill had brought a rope around his waist, and that turned out to be a tremendous help in preventing the snakes from escaping (Figure 3.3).

Other than the two big females I had at first, we captured one more very large females and three good-sized males. One of females, Mirna, the largest of them all (whom I named after Mirna Quero, my boss in Profauna), was mating, seemingly being courted by the three males. One of the other two females, Lina, was too thin to breed, and the other one was in breeding condition but I did not see males courting her. All three males seemed to have been courting Mirna. This was the first breeding ball I found, and by far the most dramatic to capture. This explains my unlikely success in hanging on to all the animals before help arrived. Females have no predators and are not worried about anybody bothering them. So, when I first grabbed them, they might have believed there was just another male coiling around their tail and saw no reason to make any serious attempt to break loose. Males, on the other hand, travel long distance and invest lots of effort to get to find a female. Being next to Mirna and having a chance to mate with her was probably the closest to a jackpot that a male anaconda is ever going to have because she was so big (and fecundity increases with size). Clearly, they were hesitant to give up their position unless they knew that there was a real problem. Because I focused on grabbing the bigger females when I was alone, they made no effort to escape, and when help arrived, it was too late for them. I was done with the males I needed for radio implantation, but Mirna and Lina were welcome additions to star in our radiotelemetry play.

We had caught all the snakes we needed for the study. Paul performed a couple of the surgeries himself so I could watch, and then he mentored and coached me while I did the rest of them. By the time Paul and Bill left, there were only a couple of transmitters left to implant, but I was well trained at the time and I managed to do the remaining ones. I had my work cut out for me: I now had to follow 12 anacondas, study their whereabouts, and learn as much as I could from these priceless animals I had marked.

Soon I learned that catching the animals and installing the transmitters was the least of my problems. Unlike other snakes whose life history is relatively well known, I knew nothing about anacondas in the wild. I did not know how far they would travel; thus, the transmitter signal had to be strong enough so that I wouldn't lose track of the animals if they moved away. The transmitter batteries have an inherent tradeoff: if the signal is too strong, it quickly drains the battery, but if it is too weak, the animal may get out of range and be lost. This is the worst that can happen in a telemetry study. We spend so much time and money wiring an animal that we definitely don't want to lose the information it may give us.

On top of this, water, vegetation, and muscle and flesh muffle the transmitter signal, and I was dealing with all three of these. If the transmitter was implanted inside a thick layer of muscle and the snake was underwater and under aquatic vegetation, the signal had a really minuscule range. Yet it was my job to stay on top of them. I made it my mission to get acquainted with my 12 new friends by hanging out with them as much as I could (Figure 3.4).

Learning the Ropes

Keeping up with the animals was relatively easy during what was left of the dry season. Radiotagged animals hardly moved, and it was easy for me to find them all in the course of one day. I spent some time keeping track of them, but I had plenty of time to work in mark-and-recapture since the dry season made new anacondas relatively easy to find (see Chapter 4). However, as the wet season came in, the savanna started to flood, and as if they were following a single mysterious command, all the snakes took off in the course of 48 hours. They were gone and out of the range of the receiver. That was when the real fun began (see Figure 3.4).

Romulo Gallegos, a famous Venezuelan novelist, specialized in writing about the people and the different landscapes of the Venezuelan countryside, painting crystal-clear descriptions of them. In his description of the llanos he wrote,

> *Tierra abierta y tendida, buena para el esfuerzo y para la hazaña, toda horizontes, como la esperanza, toda caminos, como la voluntad.*[9, p. 194]

("Open and spread-out land, good for hard work and deeds of bravery, all horizons, like the hopes, and all paths like the will" [translation by JAR])

Nowhere was that image clearer than when I had to guess the path that the anacondas had taken. In the open flooded plain, any path is as good as the next one for snake travel. Because I had not yet developed a feeling for the places they prefer, it was pretty much guesswork. I simply picked a direction where I suspected the animal might have gone and walked as far as possible in that direction, every so often trying to catch the signal with the receiver.

The first month after the beginning of the wet season was downright disheartening. All the animals had taken off, pretty much each one in a different direction. I had lost track of just about all of the animals that I was supposed to get acquainted with, save for the breeding females, which do not move much. After 12 or 14 hours of walking through the swamp or on horseback, all that I had to show for it was a high level of physical exertion, a sore butt, a hell of a suntan, and, on a good day, the location of one or two animals. Long gone were my dreams of observing them in the field and getting to know them like my dog knew me. I did all that I could to keep track of the animals I had not lost. Of course, the longer I spent trying to find a lost animal, the longer I took in keeping tabs on the animals I had not lost, and the more likely it was that I would lose some of them! To add insult to injury, because the rainy season is so rainy, I cannot count the number of times I traveled to the place where I last saw a snake to collect data on its location, but as soon as I arrived to the area the sky cracked open and a small ocean poured down from above. All that I could do was to put away the telemetry gear (which was not waterproof) and wait until the downpour stopped, if it ever did.

One day, waiting for the rain to stop, I spent quite a bit of time getting data on Monica. On my way to the site where Monica was, I had forded a river on horseback, with water all the way up to the stirrup. By the time I came back the river was a veritable torrent: There were no banks to speak of, and the current raged pretty seriously in the center. I was often given a well-seasoned horse that was not too unruly for a non-cowboy like me but plenty strong to endure a full day of work. Because he knew his way around, I often trusted him to find the way back home after the day in the swamp. I had named this one "Flower" for his habit of munching on flowers of water hyacinth at every stop in the march, and even as we marched through the swamp. I recalled one side of the river that I suspected was shallower and where the distance of strong current was considerably shorter, and I tried to ford the river through that part. Flower obediently got into the current as I bid him, but soon into it, he started veering to one side, the side we had used on the way in.

This was a seasoned horse that the cowboys regularly used for cattle herding and that knew the area well. I assumed that he knew better than I did and let him choose the path. But soon I realized that he was swimming and no longer walking on the riverbed. I was always surprised how comfortable these horses seemed to be swimming in a river with a rider on their back, but, so as not to put too many demands on him, I got off him. This was not an easy decision, since my telemetry equipment was in my backpack wrapped up in plastic bags; I didn't have a reliable fully waterproof container (remember, this is was a low-budget project, so fancy outdoor gear was out of the question). So, I tried as much as I could to keep my back out of the water. However, as I tried to get off Flower, I realized that my shoe had caught in the stirrup and he was dragging me by the foot as he swam.

I didn't want to think what would happen when we got to dry land and Flower continued to walk on the shore dragging me by one foot. Flower was nice and well behaved, but after a full day of work he was always eager to get home and would canter as soon as the path allowed it. I had lost the reins so I couldn't really control him in my current position. I struggled intensely to free my shoe from the stirrup while not getting my backpack wet but was very much afraid I was failing in the latter. When I finally freed my foot from the stirrup, I experienced short-lived relief. Both Flower and I were drifting in the current. It didn't take long for me to realize how strong it was and that I couldn't easily outswim it, certainly not carrying all my gear on my back. I saw the horse steadily fighting the current and cutting across it, moving away from me. This meant that the current was taking me but not him! I quickly doubted the wisdom of having freed myself from Flower's stirrup. Seeing that it was going to be too late, I swam furiously toward Flower, with desperate rather than powerful strokes trying to reach him, but I couldn't.

Luckily, as I swam toward Flower, my hands happened upon the lead rope that had been loosened by the current as he swam. I hung on to it for dear life and pulled my way to Flower. When I finally got a good grip on his saddle, I just let his powerful stroke tow us to dry land and focused on keeping my backpack as dry as I could, if it wasn't too late. It seemed like an eternity until I finally felt Flower's steady feet walking up the submerged bank and I knew we had made it. I reached for his reins and soon was on steady ground myself. I hugged Flower and thanked him for his strength and determination. I was just sorry I didn't have a big old carrot to reward him with, and I promised him one as soon as we got back. It took us the rest of the afternoon to get home slowly through the flooded swamp. I didn't want to look at the receiver in my backpack. I figured that it was either done for or not, and there was little I could have done with the water at my knees other than risking dropping the transmitter altogether under the water.

The most common mistake when electrical equipment gets wet is to hope so hard that nothing is wrong that you power it on. It's at that time when the circuit fries. A wet circuit will survive just fine if you let it dry first before you power it on. As soon as I got home, I pulled out the batteries of the unit and plunged the receiver into a bucket of rice. Rice absorbs moisture, and it was about the only way I had available to keep something dry in the extremely moist weather in the rainy season. I was stranded for three days. I didn't dare to power up the receiver until I was completely sure it was dry. I agonized all that time thinking of the anacondas that would get out of reach during this interruption. However, frying the transmitter would have meant the end of the study, since I didn't have funds to replace it. I used my spare time to catch up with notes and to spoil Flower. Or at least I tried. I was disappointed the next day when I brought him the big carrot I had promised him the day before. He was no more interested in the carrot than he would have been had I offered him a steak! I met similar rejection when I offered him an apple. Of course, llanos horses only feed on grass. No one ever had presented him with any sort of feed or anything other than what grows naturally in the savanna, so all the things that horses are supposed to love had no effect on him. I eventually learned that he was partial to mangos, though!

"Where Everybody Knows Your Name"

The receiver survived, and when I returned to the field, I was back in business. Luckily, after the first month of frantic dispersal, the anacondas had found a place where they wanted to spend the wet season. Just like in the dry season, they acted like "homebodies" that find a place they like and stick with it. Basically, anacondas do not like deep water. The places that hold any water during the dry season are the deepest depressions in the landscape. When the wet season begins, these places become too deep for their liking, so anacondas seek higher ground where they still have the 60 centimeters (2 feet) or less of water that they used during the dry season. So, although it seemed to me they were going far away, it turns out they were simply seeking higher ground. Once they found it, they set a relatively small home range. So, I finally had the opportunity to get to know my new friends and spend some time with them.

When the second year of the study came along, I had a lot more knowledge and knew what to expect in pretty much every aspect of the telemetry study. The second year was instrumental in giving me more data and making my ideas of the snakes' behavior more robust. The animals that taught me what anacondas were like in the first place are a bunch I will not soon forget.

Carousing Boys

The four males in the study were the ones that gave me the most trouble, not only because they were smaller and thus more difficult to implant but also because they tended to take off and get lost, to die, or to drop the transmitter during the observation period. Shortly after the implantation, I was happy to see Pablo mating with Madonna in the same place where Madonna had been found the first time, and not far from where we had caught Pablo. However, after mating, Pablo moved to a spot in the *módulo* and spent the following 2 months in the same place without moving. It all made sense to me: He had done his evolutionary mission of reproduction and was probably saving energy, not moving much, ambush hunting so I would not see him moving. Because I had a plan of not disturbing the animals for any reason, I just recorded data on his location. However, after 2 months, I needed to find out if he was OK. I found the transmitter sitting in the bottom of the *módulo*. It is uncertain if it fell out through the surgery wound, or if Pablo passed away and the rest of him decomposed before we could see anything. Pablo's lack of activity was not surprisingly, mostly because Stuart and Guillermo were behaving very much the same way. We saw them moving about for the first couple months after the surgery, but in the middle of the wet season they all stopped moving.

After seeing what had happened with Pablo, I looked up Stuart and Guillermo, only to find their transmitters also abandoned in the swamp. The hypothesis of the animals dying was becoming harder and harder to ignore. Yet the question remained: Why did they die? Paul, the veterinarian, had performed the implantation procedures on these animals, so we did not expect a problem with my newly acquired surgical skills. In fact, they survived the surgery for at least a couple of months Pablo was seen mating after the surgery, which suggested that he had not suffered any ill effects from the procedure. Was it natural mortality? Later I learned that they do suffer some mortality from year to year, but I will never know if the surgery had anything to do with it or if they simply dropped the transmitter somehow. Later some evidence showed that snakes may pass implanted transmitters via their digestive system! However, we will never know what happened to our fellows.

Three of the animals that I thought were around were dead or had dropped their units, and the other one, Gonzalo, was missing. I had no idea about Gonzalo's whereabouts until the next year when he came back to the same place where he was caught, safe and sound, but I did not obtain much knowledge of his activity and mobility. This made the telemetry of males a complete failure for the first year. Eventually, I did telemetry on more males in the later years and obtained a good picture of their mobility and habitat use. When the rainy season begins they spread out to upper grounds, traveling quite a bit over the thickest vegetation in the swamp, and through the thorniest bushes, to get to their wet season whereabouts. They set a home range that is relatively small and spend most of their time out of sight, underwater, hunting opportunistically, mostly birds, but not very many of them. When the dry season resumes males return to the place where they were in the dry season and start moving very actively looking for females for about 2 months, which is how long the mating season lasts.

Finding a Good Friend: Chuka

While I was tracking wandering male anacondas through the flooded savanna, I spent many days trying to guess the direction they had gone, traveling through the remotest corners of the swamp and trampling through all bushes and parts where I imagined they might be. I didn't learn a lot about them doing this, but it gave me the chance to spend quite a bit of time in the field, in contact with the llanos, and I ended up making an unlikely friend where I least expected it.

The wilderness in the llanos is such that domesticated animals often revert to their wild styles of living. Feral cattle, wild boars, and feral horses are easy to find in many areas. There are even packs of dogs that have reverted to their feral way of life, being able to survive for generations on the abundant game that the savanna has to offer. I have been a dog lover all my life. My last dog met an untimely death at 2 years of age due to a complication in her intestines that the vet missed. I was crushed by her loss and refused to get another dog for many years. However, the idea of seeing and learning about feral dogs had always been an appealing one to me. One day, not far from the station, I spotted a black dog with a stumpy tail who was a dead ringer for the dog I had lost. It was unusual because the ranch where I was doing my study was a private reserve and did not allow dogs on the premises—domesticated dogs, that is.

Despite this policy there was construction in the tourist station next door, and the engineer in charge brought a pet of his, a young black Lab that was allowed for only a short time during the construction, and only so long as he kept her under control. The Lab slept in his bedroom to make sure she didn't bother the wildlife. A few days later, I spotted the young black Lab playing with another black dog. At first it looked like she was playing with a mirror, and I realized that that black dog was a youngster and was coming around the houses quite a bit. I can only guess that the feral dog saw me playing with the black Lab and somehow figured out that I loved dogs.

One night I was coming back from the cafeteria to my house and heard an animal moving and shuffling in the bush not 5 meters (15 feet) away from me. I stopped and tried to figure out what it was in the pitch-black night. Being in the habit of walking at night in the bush, I didn't carry a flashlight to the cafeteria, so I had to try to use the little light from the camp to figure out what it was. Eventually a black shadow emerged from the bush and ran past me. She stopped on a dime and ran again in the other direction, circling me and doing play bows. Then I realized that it was the black feral dog I had seen, hardly visible in the dark night save for a white front paw moving frantically in the dark as she ran. I squatted and talked to her, stretching my hand toward her and trying to touch her. She stopped when she saw my gesture, and I managed to rub ever so gently the lower part of her muzzle. She reacted with great excitement and started running in ever-larger circles around me until she disappeared from view in the darkness of the night.

It was such a weird feeling: It brought back memories of my lost pooch, and I also felt excited to think that she had allowed me to pet her, as skittish as she was with people. A couple days later she did the same number. I heard the noise of shuffling feet in the bush. I knew what it was this time and immediately adopted a squatting position and asked her to approach me. I got to pet her again gently under her muzzle, only to lose her again when she ran off in the night. Next day, to my surprise, I opened the door of my house at the crack of dawn and found her sleeping on my doormat!

It was the beginning of a great friendship. She was enough of a puppy that I managed to earn her friendship and trust over time. I brought some food for her, but she wouldn't take it. The minute I turned around she grabbed a mouthful and took off, stopping more than 50 meters (54 yards) away to chew and swallow her booty. We played that game a few times that morning where I pretended to ignore food within her reach so she could "steal" it. It was about 2 weeks before she had the nerve to eat her food in front of me. She started following me whenever I went to the field to do the telemetry study, and quickly we developed a strong friendship. Some people said that I adopted her, but more accurately she adopted me since we were on her grounds. She would follow me when I went out running, or to find the snakes, on foot or on horseback, so long as the water was not too deep. She knew well what lurked under the murky waters of the llanos. When I crossed a deep river, she would stand back and return home to wait for my return at the end of the day.

I noticed that she could smell the anacondas under the water. When I approached a radiotagged anaconda in shallow water, I could see her sniffing heavily in the direction of the anaconda with her hackles way up. Clearly, she knew what was under the vegetation. It occurred to me that I could train her to use her nose to find anacondas. However, there were two problems with this idea. One was her age and rambunctiousness. Her prey drive was through the roof, which made it difficult to hold her attention with iguanas, capybara, and deer around. The other was that she had a very healthy fear of anacondas and her instinct told her to stay away from them. I didn't want to teach her to seek them because I feared that she might do so when she was on her own, and I didn't want to think of the outcome. I decided to keep her around me as my partner but to leave her natural fears intact. They had kept her lineage alive in the untamed savanna for generations, and I did not deem it wise to mess with it. Looking for anacondas with me, she still learned that it was a target we were chasing just by watching me catch and subdue them. There were many times when we were looking for snakes and she, by barking, warned me of the presence of a snake that she detected before I did. When I was trying to subdue an anaconda, her pack instinct kicked in. She would bark and jump around, distracting the snake's attention until I secured a good grip (Figure 3.5). In actuality that is when the struggle began for me, but as far as Chuka was concerned, I had secured the biting end of the snake and it was time to busy herself finding another one!

There were two "first times" I will never forget in our relationship. One was the first time I got her in a car. After we had enough of a friendship that she trusted me to pick her up, I took her with me in the truck to a site that was some 3 km (2 miles) down the road. She was a bit hesitant to get in the truck, and during the trip she looked a bit uncomfortable, but the real treat was when we got out. Chuka's confusion was very apparent when she found herself in a different place than where she had started. It was clearly a place she knew, and she likely knew how far it was from my house, so being there without "getting" there seemed to have produced quite a mystery for her.

The other time I will never forget was when I gave her a bath for the first time. Clearly, the llano is rich in biodiversity, and ticks and all sorts of ectoparasites are no exception. Once she started living in my house I had to do something about it. However, I didn't want to break the fragile trust she had by subjecting her to something that was less than friendly. One day I mustered the courage and took her to the backyard of the house that served as a field station, tied her up to a leash, and got to work. It was one of those sweltering days in the llanos, with the temperature no less than 34° or 35°C (90+°F). The perfect day to put yourself under a hose, right? Wrong! Not for Chuka. When I put the hose to her rump to test how she would react to the feeling of the cool water on her skin, I got the same reaction as if I had put a hot branding iron to her flesh. She was as terrified as if had opened the seventh gate of hell. I did not want to let her go, because I didn't want her to lose trust in me or to feel that I was someone she had to escape from. So, I hung on to her with equal resolve. We battled for no less than 20 minutes as I tried to give her a semblance of a bath while at the same time preventing her from escaping and losing her to the savanna forever. At the end of the endeavor, I had more mud on me than she had when we began and my T-shirt was torn in many pieces from her many attempts to climb over me while I tried to restrain her. I managed to get her clean enough to drag her in the house, where I could towel her down and give her abundant treats and love to let her know that I had not just turned evil, it was just something we had to do. While she eventually learned about cars and got very used to car travel in the following years, she never quite overcame her feeling toward baths!

Learning from the Girls

Relocating Large Females

The females, unlike the males, were giving me a lot more interesting information about their lives. We have already talked about Olivia, who was extremely long and very skinny at the time of capture. I saw Olivia kill a big caiman and later a big deer; she always showed the behavior of a predator on the prowl, looking for prey. Large as she was, she was one of the animals I was always excited to look for, but I also worried the most about her. She seemed to have a talent for getting herself into trouble, as she eventually did.

Some of the animals we chose to relocate. These were the ones that were very large and that we had caught relatively far away from the station. To avoid spending too much time finding them, and fearing that they could take off in the wrong direction, we moved them to a place near our site. This also gave us the opportunity to test what level of homing behavior they exhibited. Our results were conflicting. Mamacita was placed in a pond that did not have a lot of water but was right across the dirt road from a deep *módulo* where we had seen anacondas thrive. I surmised that if her pond dried out, she would simply seek the nearest water, which was 18 meters (20 yards) to the west. However, when the drought struck, Mamacita wandered east into the dry savanna, where we found her dead, likely overheated by the sun.

Mamacita was one of the largest females in the study, and I could not figure out why she had died. It would have been far easier for her to find the water of the *módulo* than to travel more than 200 meters (219 yards) into the dry savanna where she found no water and died. In retrospect, however, I could see that from the pond where we found Mamacita the first time, there was a larger pond directly to the east that held water through the year. Was she trying to find that pond located to the east when she ventured into the dry savanna, only to be caught by the morning sun too far from water? I will never know. Because we don't know what navigation mechanisms they use in their long- and short-range navigation, it is difficult to speculate.

Katalina, on the other hand, was relocated some 5 km (3 miles) from the original capture site and stayed where I had relocated her for the whole dry season, acting very much like the other animals that were not relocated. She did not try to go back and did not make any adventurous move toward her original place of capture, even though it was in the same river and I can only imagine there were chemical clues marking the way home. Instead she stayed put for the rest of the dry season, and it was not until the wet season rolled around that she, like all other non-breeding females, took off, first going down the river with the prevailing currents. Shortly afterward we found her at the first capture site, more than 5 km (3 miles) away from where I had relocated her. I guess there is an important difference between Katalina and Mamacita. Katalina was always in deep water and never had to deal with a scary drying up of her pond. Would Mamacita had been OK if I had relocated her in the deeper *módulo* just 18 meters (20 yards) to the west of the site where I put her?

One aspect that put my mind at ease is that in the same year that I lost Mamacita, and about the same week, we found several other anacondas baked by the sun, along with a white-tailed deer (*Odocoileus virginianus*), a giant ant eater (*Myrmecophaga trydactila*), and 500 head of cattle that could not survive the drought. It may well be the case that it was an extremely dry year and many animals succumbed to the drought, Mamacita being one of them. I will never know for sure, of course.

Ethical Considerations

The death of Mamacita disturbed me. Did my actions have something to do with her death? Sure, I learned something for science thanks to her. I dissected her body and obtained valuable information about reproductive biology, diet, stored fat deposits, and other data that one can only get from a dead animal. However, that was not the kind of science I wanted to do. We can find the whole gamut of positions about what is acceptable to do to an animal in the name of science. I do understand the need to sacrifice animals in the name of science. I have done it on occasion when traveling to remote locations where the diversity is unknown and the specimens are important for museum collections. However, I would much rather work with live animals. To put oneself in the animal's shoes is a good approach because it teaches you things about their behavior you would not be able to appreciate otherwise. But once you have done that, you establish an empathy with the organism you are studying that makes it very difficult to ignore their pain or to be insensitive to the passing of a study animal.

Every scientist needs to develop his or her very own ethical standard of how to treat animals, and I truly believe that there is no one that is better than any other. It is just what the scientist feels is the acceptable standard for himself or herself. I have chosen an anthropocentric standard: If it is OK to do it to a human, it is OK for me to do it to an animal. For instance, catching and forcibly restraining a person is acceptable. This is what cops may do to a suspect; it is acceptable so long as they do not use unnecessary force. This is the approach I use: say, subduing an anaconda, holding it, pinning it, and applying only the amount of force needed to subdue it and no more. When I studied the diet of iguanas, the standard method to study lizard diet was to kill them and look at the stomach content. This was the standard procedure, certainly in the 1980s, when dealing with non-endangered animals. I could not get myself to do it, but a person who has swallowed something toxic will be subjected to a stomach wash to remove the stomach contents. It is not a pleasant thing: They pump water inside the person's stomach and squeeze it out with external pressure. They may repeat that treatment several times until the toxic substance has been removed. Since it was acceptable to do to a person, I was willing to do to an iguana, in the name of science. I ended up doing more than 300 stomach washes to understand iguanas' feeding behavior but did not kill any specimen for this study.[11]

I have taken this standard seriously. The accepted way to mark lizards, at least at the time before pit tags and other fancy methods came along, was toe clipping: You would clip one or two toes, assigning a number following a code. When I was presented with that possibility of marking iguanas I asked myself if I would like to have a finger or a toe clipped off in the name of science. I clearly didn't think that science was important enough to give up one or two of my phalanges, so with what authority could I say that the iguana had to do it? Why would the iguana make a sacrifice for science I was not willing to make? And how could I do something to an iguana that I was not willing to do to myself?

Another possibility that was presented to me was branding them. Cold branding, involving liquid nitrogen, is reportedly painless, but with my Venezuelan budget, liquid nitrogen was out of the question. How about hot branding? It was far less complicated logistically. I could make the numbers 0 to 9 with paperclips and pliers, attach them to a little wooden handle, and use a small camping stove or lighter to heat it. But is that something that's acceptable to do to a person? Hardly! Was marking the animals important enough to me that I was willing to endure that treatment? It was certainly better than losing a phalange but still did not look appealing. How painful was it really? There was only one way to find out. I made a wire number, heated it up on a stove until it was red, put it to my forearm, and counted. I held the number to my arm for 1.75 endless seconds. It turns out that on the delicate skin of an iguana, a split second of contact with the hot wire produces a neat mark that does not penetrate the skin and thus does not risk infection of the wound. This marking can be read years afterward. I would not claim that the branding didn't hurt the iguana or that I know how much it hurt the iguana, but I least I felt that I had the moral strength to impose that treatment on the iguana. It was important enough for science that I had done it to myself, so I felt it was OK for me to do it to an iguana.

So, over the last 25-plus years, I have certainly caused stress and pain to many anacondas. I have done surgeries, cutting them open to implant devices inside them. These procedures are acceptable for a person. I have done these procedures using the proper anesthesia and proper sterile techniques, and I have tried to make sure to lower distress and pain for the study animals at all moments. In the process of capturing and handling anacondas I have lost some animals. Certainly, when I was learning the ropes I did not fully understand what can risk the life of an animal and what does not. I wish I could say that I will not ever lose another study animal, but I *can* say that I always do my best to lower pain, distress, and risk to the health of the animal. That makes all the difference!

Where Do Anacondas Live?

Since I had put so much effort into catching the animals and doing the surgeries, and had subjected them to such treatment, I owed it to them to obtain as much information as I could from them. To get the most out of their sacrifice, I set up a rigorous observation regime. Every other week, I set up a regular schedule in which I would locate the animals every 2 hours for a 24-hour period in order to get a good idea of their activity patterns. These were excruciatingly tiresome undertakings but gave me the opportunity to learn the most about them. I started it during the dry season when the animals were around and had not moved much. In those days, I could observe them all during this 24-hour period because it didn't take long to get to each animal. After 24 hours running around finding snakes without pause, I was ready for a good night's sleep, but it provided me with valuable information. Of course, I had to cut my losses during the wet season and work only with the snakes that I could get to, and eventually I had to stop altogether when the animals moved too far away (Figure 3.6).

While making these observations, I had the opportunity to dispel the myth of nocturnal behavior of the anacondas. Since I was seeing them day and night, I could see when they moved the most. They don't move a lot during the heat of the day, preferring to move out on land when the temperature is cooler. Their activity during the night is no different from what they do when the sun is suddenly covered by clouds. They are just as likely to come out and move during the day at, say, 4:30 p.m. if the sun is covered by clouds as they are at night after the sun goes down. If there is any pattern, there is higher activity at dusk and after 2 a.m. there is hardly any activity at all, so the idea of nocturnal habits does not seem to be supported. Activity seems to be associated more with a temperature bracket at the end of the day.

It was during one of these observations when I found Susana, a big female we were monitoring. I saw one of her coils breaking the surface in the 4 a.m. observation. I surmised that she had caught a prey and was constricting it under the water. I was eager to make the 6 a.m. observation because I'd be able to see with better light. Whatever it was likely would not be swallowed too quickly, so I expected to observe it in the morning. However, when I saw her again at the crack of dawn, she had not moved. For a minute I entertained the possibility that she might still be constricting a prey—killing an aquatic turtle under water may take hours—but soon I realized that she was far too still for far too long. I touched her with a pole and she bobbed on the surface. She was dead, beheaded likely by a capybara defending her offspring (see Chapter 5).

Trying to follow anacondas year-round, retracing their steps (so to speak), gives you a new perspective about our differences. They can glide so effortlessly through thorny bushes and under a carpet of aquatic vegetation that when you try to follow them to better understand their life, you get a good reminder of how different we are. Getting into a pond, I could do a personal evaluation of how much I like that pond. If the answer was "not at all," I was almost certain that it was where the snake was. If it was stagnant water, that was good; if it was surrounded by thorny bushes, that was better; and if the water festered with noxious smells, that was best!

Anacondas like caves. In the rivers with forested banks, the roots of the trees often hold the bank from being eroded altogether, but under the roots of the trees sometimes there are caves that persist because the roots prevent the bank from collapsing over them. These tend to be great places for aquatic organisms to hide. Caimans, iguanas, tegus, and toads often share these caves with the anacondas. For some reason despite their regular trophic relationships, I often found caimans sharing the caves with anacondas big enough to eat them. In both dry and wet seasons, I found anacondas using these caves. In fact, the absence of caves seems to be the reason that there are fewer anacondas in the parts of Guaratarito that have no trees. Interestingly, the proportion of males and females found in Guaratarito was strongly biased toward females. It seems to be a place where large snakes come to feed but does not seem to be a good place to find many males.

Expecting Girls Just Chill and Bask

One thing that became apparent quickly as I was trying not to lose the animals was that females that were breeding did not go very far. They dispersed from their dry season hangouts, to be sure, but soon found a place in the river bank, or right by the road, and stayed there the rest of their pregnancy, basking intensively but not moving much. Mirna was the largest female in the study; she was thick and massive. I was excited to see her giving birth. When the season came, I was on top of her because I was really interested in learning about any parental care or behavior of the neonates. Clearly, such a large snake would deter any possible predator (foxes, herons, and such) from preying on her offspring. One day Mirna, who lived in the same bend of the river for the whole pregnant season, was gone. I followed her signal down the river up to a cave, where I found her substantially skinnier than she had been the day before.

I knew then that she had given birth, and I tried to find the babies in the area. However, I failed to find a single one of them, no matter how much I scoured the riverbank, nearby vegetation, and river caves looking for them. I was tracking two other females at the time and doubled my efforts to find those giving birth, to no avail. I calculated I may have missed some 200 neonates that slipped right under my nose and into the swamp without me noticing them!

Bachelorettes Have the Munchies

Non-breeding females were the ones who gave me hardest time. They moved longer distances and once in a place continue to wander and prowl around. This was the case of Olivia, as I mentioned, and Lina, who was found near a breeding ball that Mirna was hosting but was not in breeding condition herself. As I learned later, she was likely waiting to dine on some of Mirna's boyfriends. They were the most fun to track. They were the more likely to be found in the act of killing a prey or digesting a meal. Following them, I learned a great deal about their diet because, unlike their breeding counterparts, non-breeding females are in the business of gathering energy to reproduce the next time around, and they forage very actively (see Chapter 5). After they stopped moving for the wet season, it was very enjoyable to go to the place where I had found them and try to guess where they were. Eventually, I could arrive at a place and quickly spot the likely area where they were based on vegetation, water depth and coverage, and other microhabitat features before I even turned on the receiver, which I used only to confirm my suspicions.

Thermal Biology

In the study of reptiles, the issue of thermoregulation often plays a central role. Unlike mammals or birds, non-avian reptiles do not produce the heat they need for their metabolic functions, so they need to obtain the heat from the environment. Snakes are especially gifted in this area since they can change their body shape more than other reptiles. Coiled snakes have much less relative surface area than stretched-out ones and exchange heat with the environment far more slowly. A snake that is too cold can stretch itself out and obtain heat from the environment more quickly. On the other hand, if it is at the desired temperature, it can coil tightly and reduce the amount of heat transfer, remaining at its desired temperature. Water gives and takes heat a lot faster than air, so anacondas may also use their position in or out of the water to regulate their heat balance.

My observations on anacondas' thermal behavior are consistent with this notion. Anacondas that need heat lay stretched out in the early morning sun, which helps them raise their body temperature. It is possible to see them coiled up in the shade or a cave toward the middle of the day when presumably they have acquired the desired temperature and the baking sun of the llanos would overheat them. However, the temperature of the llanos is normally warm enough, around 30°C (86°F), that anacondas seldom have to go out of their way for thermoregulation. Olivia had a temperature-sensitive transmitter, and her temperature often was between 27° and 29°C (80° and 84°F), a few degrees under the 30° to 37°C (86° to 98°F) that the air and water reached during the day, but she managed to keep the same temperature after sundown when the air and water temperatures dropped to between 22° and 25°C (71° and 77°F). While she managed to keep a more stable temperature than the surrounding environment, it does not seem like her thermic homeostasis abilities were regularly very challenged in the constant warmth of the llanos.

Conventional wisdom about the thermal physiology of snakes, which was developed for northern and temperate snakes, suggests that they bask after a meal, which increases their metabolism and helps them digest it. I did not find any evidence of this in anacondas that had eaten a recent meal. Instead, they stay in the water, where the lump of the ingested meal often floats. This spot may be exposed to the sun and presumably becomes warmer, but on occasion a patch of aquatic vegetation covers it, shielding it from the sun. It seems like it is more an accident due to the gasses produced by digestion and decomposition than a deliberate behavior to increase the temperature of that part of their body. Again, the warmth of the area releases the anacondas from the thermoregulation needs that temperate snakes have.

There were two situations in which I did see clear evidence of anacondas deliberately basking in the sun, seeking heat. One situation is when a female is pregnant or about to become pregnant. Females that are near the breeding season lay in the sun for hours. They may be stretched out completely or folded once, basking near the shore, seemingly raising their body temperature. Near noon they may go in the water, but they still stay near the surface in very shallow areas, where the water heats up substantially during the day.

The other situation in which anacondas may bask for long times or seek very warm bodies of water is when they are wounded. Olivia received a bad injury in the nape, near the head, likely the bite of a capybara, that left a gaping wound in her neck. Not knowing if she was going to survive, I followed her carefully and found her in a shallow borrow pit that was 35°C. According to her transmitter her internal temperature was 36.6°C! Olivia stayed in this borrow pit for 2 more days. I surmised that she was giving herself a fever by basking and spending time in a very warm pond since hyperthermia is a well-known strategy to fight microbial diseases. Over the course of my study I have found no shortage of basking anacondas, and with hardly any exception, they were either breeding or wounded.

4

Anacondas' Tales

Sickness and Health, Birth and Death in the Life of a Giant Snake

Cuentan que una hermosa doncella llamada F'rimene hermana de Nuna (la luna) huía por la selva escapando de la ira de Wanadi (representante celestial en la tierra). Durante su huida se tropezó con el rio Uri'ñaku (Orinoco) que junto con los demás ríos, acababa de nacer. Como no podía pasar el gran rio se dijo a sí misma: El agua va a ser mi camino, soy la dueña del agua, madre de los ríos y se convirtió en Hui'io la Gran Serpiente que hizo su casa en el fondo del rio.

(It is said that a beautiful maiden called F'rimene sister of Nuna (the moon) fled through the forest escaping the wrath of Wanadi (celestial representative on earth). During her escape she ran into the river Uri'ñaku (Orinoco) that along with the other rivers had just been born. As she could not pass the great river she told herself: Water will be my path, I am the master of the water, mother of the rivers and turned into Hui'io the Great Serpent that made her home in the bottom of the river.)

Makiritare legend[12] (Translated by JAR)

A common theme among snake biologists is the bad PR that our beloved animals have. Few animals are as loathed or feared as snakes are. This loathing/fear seems to scale up with the size of the snake. Large snakes are especially targets for elimination. Part of this may be an evolutionary response from millions of years, in Africa, coexisting with large snakes as predators of primates. Because of this the life of snakes is not an easy one, but their enemies are not limited to humans. Even the venomous and giant snakes have many enemies. Their body constitution makes them very vulnerable. Their skull is fragile and easily wounded; they do not have limbs to protect themselves or to deter an enemy with claws. They are not fast runners. Their dentition is not very impressive or showy, even if that of vipers may be deadly. What good is it to kill your enemy if you die of wounds inflicted by it? For most snakes the only ways to avoid damage are not being detected, escaping quickly, and warning potential enemies of their danger with different displays. The case of anacondas is especially bad. A large female is certainly an imposing animal, but even she can suffer death from a number of causes, as we will see. Smaller anacondas have a very raw deal and suffer high predation.

In this chapter I will share what I have learned about their lives and tribulations. I use a long-term dataset, opportunistic information gathered in the field, anecdotal encounters, tales people told me, and my gut feeling to try to paint a picture of what it is like to be an anaconda.

Population Sampling

Population sampling techniques are not always driven by the formal scientific method. Often gut feelings, intuition, accurate observations, and sheer luck have a lot to do with how we conduct a scientific project. As I mentioned earlier, I learned a lot of the sampling techniques from the day-to-day knowledge of local workers.

Of the cowboys who taught me so much about the llanos, perhaps among the most memorable was Don Pedro. He had lost his eyesight to glaucoma years earlier. Being blind, he had retired to a low-impact job watering the ranch's ornamental plants during the dry season. I was always surprised to his ability to find all the ornamental trees that needed watering without being able to see them. Don Pedro knew so much about the llanos that when I could not recognize a plant species, I brought specimens to him. He would identify them just by the feel of the foliage and the smell of the leaves!

Once Don Pedro walked into the garden of the house where I lived, and for no apparent reason, stopped in his tracks and asked me if there was a cat in the bushes in front of him. To my surprise, I did find a cat hiding on the opposite side of the bush! The rascal seemed to be sneaking up on a bird nest. Amazed, I asked him how he could possibly have known about the cat. It was difficult for even my young eyes to spot the hidden feline in the bushes. He replied that it was because of the frantic calls of a pair of wrens (*Campylorhynchus griseus*). He told me that these wrens make that particular call only when they see a cat or a snake. In the past, he told me, he had avoided unwanted encounters with rattlesnakes by exploiting the keen eyes and attending voice of vigilant birds.

Many birds feel very strongly about snakes. They can't help themselves, seemingly compelled to flit about the predator and call noisily. This distinctive behavior has been called the secondary predator attraction hypothesis. Smaller birds call to draw in other predators (e.g., large raptors, canids, felids, crocodilians) to feed on or drive off their nemesis, the snake. Both the small birds and the snake-eating secondary predators that attend to them benefit from this co-evolutionary arrangement. Be that the reason or not, the fact remains that birds feel very strongly about snakes and always feel the need to mob them and make a big racket when they see one.

This is especially true with wading birds like the wattled jacana (*Jacana jacana*), which are prime prey for juvenile anacondas. They often mob and vocalize characteristically when they see a snake in the swamp. I did not have the trained hearing of Don Pedro, but, with practice, I learned to recognize the calls of my previously underappreciated accomplices and was able to greatly increase our success at finding snakes.

Other than those times when the avifauna of the llanos helped give away a snake, the only way to find anacondas was to get down and dirty. We would slog through the muck and vegetation, searching by feeling with poles and feet. During the intensive searches, I had plenty of time to try to put myself in the snake's shoes and fully appreciate both my "unsnakiness" and the limits of my imagination. Unlike me, snakes would not have to endure heat, would not feel hungry or thirsty, and would not feel fatigue as I felt forcing myself (and my crew) to trudge through the marsh for hours on end. Unlike human bodies, snake bodies are custom built to glide effortlessly through the vegetation without making so much as a ripple. Their supple maneuverability and tough outer covering (imagine your entire body covered in row after row of overlapping fingernails) allow anacondas worry-free movement through the thorn bushes and cutgrass that torment human waders of the llanos. I, thus, had ample time to admire and appreciate the features that make anacondas the masters of the swamp. This gaze into the abyss that separates our worlds would prove to be as deep as the challenge to fully understand their biology.

Mark and Recapture

The core of the demographic study was based on mark and recapture. The idea is that captured individuals are measured and marked before they are released in the place of capture. In later years the animal may be recaptured, and after we collect its measurements and identify it properly, it will yield information on growth, mobility, survival, and population size. The method relies on marking many animals and hoping to recapture enough of them to have statistical representation on information about survival, growth, and mobility. It looks fine on paper when you want to study animals that you can trap in conventional traps or count in conventional transects. Such techniques are simple and straightforward. However, when it comes to slogging through the swamp trudging on vegetation, submerged logs, and caimans and hoping to run into an anaconda, the situation takes a completely different twist because you may spend hours upon hours looking for the target animal. Then you need to capture and subdue it (Figure 4.1). Figure 4.2 shows a summary of the sizes of snake captured in the study. The largest snakes I encountered are only a bit larger than 5 meters long (almost 18 feet).

I will never forget my first encounter with a large female anaconda in the wild. I was in Masaguaral, Venezuela, well before I started my formal work

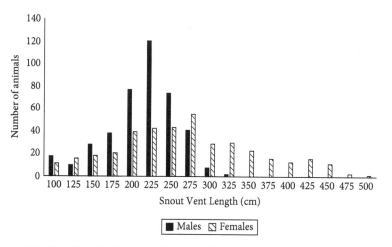

Figure 4.2 Size distribution of wild anacondas. Notice peak of animals between 2.5 and 3 meters (8.2 and 9.8 feet). This corresponds to the large number of males that stop growing at that size. This figure shows one of the first things that struck me about anaconda biology: Males do not grow very large and only females reach the very large sizes. This is called sexual size dimorphism, when the two sexes are of substantially different sizes (dimorphic), and became a major part of my later research (see Chapter 6).

with anacondas. John Thorbjarnarson, my friend and mentor, was doing his Ph.D. dissertation with caimans, and I was working on my undergraduate thesis with iguanas. We were hanging out with a couple of other friends around lunchtime when someone, knowing we were herpetologists, arrived with the news that they had seen a big anaconda basking next to a bridge on a neighboring ranch. They insisted that the snake was not moving and we could likely find her if we went looking for her.

We were so excited to hear this that we didn't pay much attention to the other part of the news—that they had seen the animal the day before. Had we taken that into account, we wouldn't have made the trip. Of course, then, we didn't know that pregnant females are the only ones who bask and when they do, they do so very regularly and in the same spot for most of the season. But as fate would have it, we went to look for the snake.

On arrival we split into two groups. One had Walter Timmerman, who was visiting John from the United States. The other had Kako, a hired helper who assisted John in maintenance of the crocodile farm in Masaguaral, and myself. After a few minutes searching, I heard a distress call from John. They had found and engaged the snake but were outmatched by its strength. John feared they would not be able to hang on to it. Kako and I joined in trying to pull the snake, but she anchored herself on the column that supported the bridge. Slowly but surely she was dragging us with her to the water.

As the snake was slipping from our fingers, I had the oddest sensation. It was as if I simply knew what to do in that situation, as if the knowledge had surfaced in my mind when presented with the real problem and I simply was remembering it. I reached into the water at thigh level to undo the anchor the snake was using as leverage to beat us. It was very simple physics. Immediately she started to give way. She made another anchor with another part of her body in another column of the bridge, which I undid the same way. As John and the others were now gaining on the snake and pulling her out of the water, by now I was very close to her head. I knew this despite the murky water because I could feel the narrower girth of the part I was holding. It was not long before I felt the fast movement of the snake's head thrashing around in the water seeking to bite her aggressors. I jumped back and instinctively reached into the water where I predicted the snake's neck would be. Jackpot! The wide-open mouth of the snake twisting herself in my hand told me I had been successful and had secured the head of the giant.

Of course, this was only the beginning of the struggle. The snake became really ornery, twisting her powerful loops around my arms, trying to protect her head, and making it very difficult to hang on to her. I had managed to thwart her plans of escape, but we were far from gaining control of her. I knew the last thing I wanted to do was to let go of her head, so I used all my strength to control that part while John and the others dealt with the rest of her. After I learned about the "evil loop" I realized that I did exactly the right thing at that time. The four of us struggled for a good 20 minutes before we could claim to have controlled her and were able to put her on the truck to take her to the station for measuring and processing (Figure 4.3).

Finding Snakes

As a graduate student at the University of Tennessee, on the third year of my project, I wrote a grant request seeking a small amount of funding for travel and field expenses. In the "Methods" section I described how I go about looking for snakes in the swamp. I will never forget the comments of one of the reviewers (I am paraphrasing): "Plans to wade through the flooded swamp hoping to trip into aggregations of mating anacondas. I can't imagine few projects less likely to succeed." Clearly, that grant request was not funded, but it provides some context for why a project like this had not been attempted before.

Over the years I learned that the odds of finding a snake vary tremendously depending on the level of flooding of the savanna. Water levels may vary due to differences in the amount of rain that fell during the previous year and due to the management of water in the *módulos* by ranchers. When the savanna was flooded, the only way we could search for snakes was from trucks on the raised roads. We could use boats, looking for basking animals or those crossing the roads. Intensive searches are much more effective during the dry season, whereas cruising is a better method during the wet season and transitional periods. The relatively high number of animals found during the height of the wet season largely represents pregnant females basking conspicuously. We did intensive searches, on average, 55 days of every season. In 6 years of intensive demographic sampling (1992–1997), we caught an average of 125 snakes per season. This yields an average capture of only 2.27 snakes a day in intensive searches even during the high-capture days. In later years, when we focused our sampling on drier times, we obtained a higher capture rate, but it always depended on the seasonality.

Let me put this encounter rate in context. We needed about three or four people working 8 to 10 hours a day in order to catch a little more than two snakes per day. This is great news for biologists and snake junkies. Anacondas are so elusive and so difficult to find that I would not have been so worried about the success of the project had I known I would have this rate of success. That reviewer would have been a lot kinder if he or she had known the odds of capture. However, it is one thing to look for anacondas for study and quite another to look for anacondas for profit in a commercial operation. When you calculate the labor needed per square foot of snakeskin, the commercial cost of this operation would simply be prohibitive.

A conservationist can take this piece of news in two ways. First, because the cost is so high, a program such as this would not be sustainable and would never happen, so there would be no need to worry about it. Another, more worrisome way of looking at this is that the cost would be so high that administrative authorities would have all kinds of incentives to "cut corners." Tricks often included intentionally overestimating populations, hunting in forbidden areas, ignoring assigned quotas, and targeting individuals, including culling gravid females. It seemed it would be next to impossible to design a management plan that both respects the biology of the animal in accordance with sustainability principles and satisfies the economic needs of the market. Despite all claims to the contrary, I knew that the true purpose of studying anacondas was not really to see if management was possible but to give an appearance of legitimacy to a program they had already decided to do. Otherwise, they would not have committed more than $30,000 to this enterprise. What I was learning was that harvesting anacondas properly was very difficult because it would result in management that is not in compliance with the biology of the species, or with common sense for that matter. Knowing how things worked within Profauna and the subservient attitude to big businesses that the Venezuelan government had in the early 1990s, I could easily foresee a disastrous and exploitative "management" plan to harvest anacondas that would seriously threaten the local populations.[2]

One of the problems with starting this project was that Profauna was an administrative institution. As part of the packages and adjustments that the Venezuelan government had made in the past decade to qualify for international aid (from the International Monetary Fund and World Bank) Profauna was instructed to stop doing research itself and instead outsource it to groups like universities and other research institutions. The idea was that the government would not fund research anymore; rather, the private sector would sponsor the research the country needs. Of course, it was also limited to the research that the private sector was interested in. So, this requirement to give up doing research in reality dismantled the country's infrastructure and made the country technologically dependent.[2] As a result, in the early 1990s Profauna did not have any research equipment or staff researchers or any idea of what research was. So, my boss had the idea that I would be in the field studying anacondas by myself, without any support, help, or even a vehicle! That was what the budget allowed for.

Fortunately, John Thorbjarnarson (who went by the nickname Juan Caimán, aware that his name was quite a challenge to pronounce), through the Wildlife Conservation Society, had secured not only some extra funds, a vehicle, and a salary for a helper (María Muñoz) but also some important collaboration from key people like Bill and Paul. Bill not only came the first few years to help catch snakes but also had contacts with a network of enthusiastic and knowledgeable US zookeepers who came to the field to help me with the fieldwork of catching and handling the snakes. I am very thankful for all the people who came over the years, spending their own money and their own time to help with my project. For them it was a special adventurous vacation, and for me it was invaluable help with handling the animals, especially at first before I had figured out all the tricks to dealing with the evil loop and how to handle big ones.

Among the people who came was a large bodybuilder one of my friends nicknamed "the Lummox." He was a friend of a zookeeper and a snake lover but didn't have much field experience—or any outdoor experience, for that matter. But he was strong and a big snake enthusiast, so he was always the first one diving to catch the big snakes. His help was very much welcome given the hardships of the fieldwork. However, the Lummox did not seem fully acquainted with the joking nature of the Venezuelans or the ruthless Venezuelan sense of humor. Because we were in the field and he was not very knowledgeable about the field and wildlife, he was gullible and often targeted for jokes by me and the other members of my crew.

There is a bird, I believe some sort of ibis, that walks on the mud of the swamp, pecking into it and feeding on small invertebrates. As it walks, it leaves a trail of holes on either side of its path. These holes are about 40 centimeters (1.5 feet) apart, as the bird pecks on one side or the other, and they go on for a long distance, until the bird stops foraging and flies away, at which point the trail disappears. When the Lummox asked about this mysterious trail, one of my friends told him it was a giant Amazon centipede that probably had dived right where the trail disappeared. It was something to see his face imagining a centipede that was capable of leaving a track 40 centimeters (1.5 feet) wide and that was likely submerged 2 meters (6 feet) from where we were standing!

Later on, he made the mistake of asking if there were leeches in the water because leeches were the one thing he absolutely could not deal with. The honest-to-God answer is "yes." Every time we went into a swamp, it wasn't uncommon for one or two crew members to come out with one or two leeches attached to their feet. This isn't a big deal: You only need to yank the leech out and toss it back in the swamp or on the sand, depending if you want to kill it or not. The only consequence is that, as their saliva contains anticoagulants, the small wound where the leech was feeding produces a small trickle of blood for a little while until the anticoagulant is washed away and the blood can coagulate.

However, as the Lummox acknowledged his weakness, my friend Cesar Molina was quick to tell him that there were indeed leeches, some as large as 30 centimeters (1 foot) long. To add to his terrified reaction, I chimed in, saying that it wasn't so much their size as the fact that they form big schools. Cesar doubled down, telling him how he had seen them taking down a puma that was trying to cross the river and never made it to the other side!

The Lummox eventually figured out that we were not serious and mustered the courage to get in the water. Naturally, some of us ended up with leeches attached to our feet. Cesar showed one to the Lummox and when he pulled it off, instead of tossing it back in the swamp, he was careful that the Lummox saw him place it between wet napkins to keep it alive. The rest of the week Cesar would chase the Lummox around with the leech, the Lummox emitting high-pitched screams as he fled. Sometimes the Lummox would stand his ground holding with a wooden stick, vowing to knock Cesar's head off if he came any closer, only to lose his nerve when he realized that Cesar could still toss the leech at him. Then he'd resume running away in panic. Sometimes all Cesar needed to do was to chase him holding a wet napkin that did not contain a leech. All good fun!

Of course, there was a serious part to the study. The animals we caught were processed and measured thoroughly in order to obtain information from them. Whenever I caught a snake, I marked it by scale clipping. Clipped scales may regrow, but for a long time they show the darker color where the scales were cut. I also copied the drawing of the subcaudal ventral spots on the first 15 scales (digital photography was too expensive at the time for me).

These spots do not change during the snake's life and provided a backup that allowed me to confirm the snake's identity even after the clipped scales had regrown completely. I could recognize snakes using this marking even 13 years after the original capture.

How Many Anacondas Are There?

A common question people ask me is about the population size of anacondas. How many anacondas are there in my study site? How many are there left in South America? At first, I was in no position to answer this question, but as I continued doing the mark-and-recapture study, a picture emerged about anaconda abundance. At the time of this writing, I have caught 917 animals (473 males and 444 females) in an area of approximately 2,500 hectares (5,500 acres). Dividing the number of animals caught by the surface they inhabit results in a density of 0.37 animals per hectare (917 individuals/2,500 hectares). But I knew I had not caught all the anacondas that lived in this area. A simple method of population estimation (Lincoln-Peterson) consists of marking a few animals and letting them go. Then we make another capture effort, where some animals will be marked and others not. The proportion of marked animals in the second effort is considered the total proportion of marked animals in the population. So, since we know the number of animals marked originally and we know that in any given year a bit more than 20% of the animals were marked, the total population size may be easy to calculate. In other words, the 917 snakes I had caught represented one-fifth of the total number of anacondas present. This calculation produced the staggering number of 4,585 anacondas in my study site, at a density of 1.8 anacondas per hectare!

A density of 0.37 anacondas per hectare seemed high to me, but the 1.8 figure was just too much. Sure, anacondas are abundant in this part of the swamp—but that abundant? Anacondas are top predators, of large-sized prey, and are strict carnivores. Could they really find enough nourishment in such high density? On the one hand, anacondas are ectotherms, meaning that they don't spend energy in heat production; rather, they get their heat from the environment. This is not such a bad idea in a place where the average temperature is around 30°C (86°F). Why spend energy producing heat when there is so much heat in the environment? But I still had trouble coming to grips with such a large density of a top predator.

Figure 4.2 shows the animals that we found over 25-plus years and not necessarily the population at a given time. The simple calculation I made for population size refers to a closed population that does not grow or where marked
animals do not die. So, over the span of more than 25 years that I was using,
there were bound to be deaths, births, migrations, and so forth. Therefore,
I needed to use more sophisticated models to help me find out what the population size really was at a given time. More advanced mathematical models
using recapture of marked individuals over several years allow for a more
accurate calculation of population size and can provide estimates of mortality, survival, encounter rate, and other population parameters. The results
of this model were rather surprising: The population comprised 286 males
and 108 females. This lower calculated abundance yielded a population density (0.16) that was more reasonable for a top predator. However, this was
still quite a high value for a top predator that needs a base of prey to live
on, ectotherm as it may be. This high density of a predacious carnivore may
be possible because marshes are very productive habitats. They have lots of
vegetation, lots of water, and lots of sun. The llano is a new deposition that
received sediments from the Andes that are rich in nutrients. This represents
a large base for the food chain. Thus, the density found, although impressive,
at least is not unrealistic.

Beyond the population size, I was taken aback by the model estimates, in
particular by the calculated sex ratio. The model output presented me with
a number of new questions. Scientists always look for answers to questions
we have, but we are even more excited when instead of finding the answers
we find new, more interesting questions to tackle. During the fieldwork it
was obvious to me that I caught more males than females (1 female for every
1.13 males), but I felt that this was due to the males' higher mobility and that
the sex ratio in the field was not really different. If the males move more, say,
looking for females in the mating season, I was bound to find them more
often than females even if their real abundance was pretty much the same.
However, the model calculated that the sex ratio was 2.65 males per every female. Why did the capture rate not accurately represent the sex ratio?

I must confess to having a conflict about this disagreement between
the field data and the results of the model. I view myself as "Mr. Real-Life
Biologist." I always felt that mathematical modeling is a tool that is out there
when needed, but the real data should take precedence over the results of
models. I have always felt that if the results of a model don't match the real
data, we should throw away the results of the model and we should keep the

real data. Based on the information I had collected in the swamp during my everyday work, there was a similar sex ratio, but according to the information from the mathematical model, there was a strongly male-biased sex ratio. I needed to explore further if there was any truth to the model's findings. Could the collection of field data be biased in the direction the model indicated? If the true sex ratio is strongly male-biased, my data collection needed to have a strong bias toward detection of females to give me a nearly even sex ratio in the animals gathered from the field.

Reflecting on the fieldwork, I concluded that, if anything, my sampling might be biased toward finding more males, not more females. During the mating season, males move about a lot more looking for females. Thus, they're easier to find moving across roads or traveling over dry land. Breeding females, on the other hand, hardly move during the mating season. They stay in one place to be found by males. As I will discuss later, only a fraction of the females breed in a given year. The females that do not breed hide under the vegetation ambushing prey and are, if anything, more difficult to find. To my mind, these considerations explained the small male bias in the sampling and led me to think that the actual sex ratio was in fact even, as I know it to be at birth (see Chapter 6). On the other hand, if the results of the model were to be believed, my sampling was drastically overrepresenting the females. Could this be?

Exploring the possibility that females may be overrepresented in my sampling, we find that females are, after all, much larger, which would make them easier to find under the vegetation. To grow larger, females need to eat more and more often and eat larger game (see Chapter 5). When a female is handling or digesting prey, she becomes easier to spot because her body breaks the surface of the shallow water (Figure 4.4). Feeding on larger prey involves longer handling times (killing and consuming) and becoming more conspicuous after ingesting the prey. This would render females easier to spot by our crews. Finally, females that are breeding are in breeding balls that are more conspicuous than an ambushing anaconda and bask very often, making them easier to find. Males, in contrast, have many potential predators and would benefit from being secretive. All these factors would bias the sampling toward detection of females, as the model results indicated. Is the bias toward female detection for their large size strong enough to counter and exceed the bias in male detection for their higher mobility? I was left to ponder if this is reason enough to adopt the results of the model over my raw data from 25-plus years. Not an easy choice.

Mortality and Survival

Beyond my preference for real-life data over estimates from models, I had another problem that prevented me from accepting the results of the model. I knew the sex ratio at birth was 1:1 because I kept pregnant females in captivity until they gave birth (see Chapter 6). Anacondas, like humans, produce the same number of males and females because the sex ratio is determined by sex chromosomes. If the sexes start out equally abundant, why would there be so many more males out there in the wild? In some vertebrates one sex or the other disperses away from the area they were born. In mammals it is often the juvenile males that leave the natal homeland, but the trend in reptiles is not clear. I do not believe migration plays a significant role with anacondas. For starters, males are the sex with the wanderlust, not females. If an individual would move out of the plot, it would more likely be a male. So, this would not account for the lower number of females. Furthermore, the movements into the study site must be balanced by movement out of the study site, so there would not be a change in the net number of animals. Finally, migration should not be very high due to the fact that the study site is the only region that holds water: Thus, animals might move out in the wet season, but they would likely come back in the dry season, when I did most of the sampling.

If migration is not the answer for the uneven sex ratio, we are left with mortality of females as the only explanation. Of course, if females suffered higher death rates, that could explain the male bias in the population that the model estimated. At first glance, this comes as a surprise because one sees males being preyed on relatively often but not females. The range of predators for male-sized anacondas far exceeds that of predators for female-sized anacondas. Predators for small anacondas are many and varied, including birds of prey, caimans (Figure 4.5a), foxes, other carnivores (including cats; Figure 4.5b), and even anacondas themselves that on occasion might engage in sexual cannibalism. On the other hand, the number of predators that can attack a female seems lower. So why would females have higher mortality than males? And how much higher is it?

Looking further into the results of the population model, I found that females have a lower estimated survival than males. While males have a 77% estimated survival from year to year, females have a probability of survival of 64%. This internal consistency in the results of the model gave me reason to take it more seriously. Regardless of the reasons, if females do indeed have higher mortality, it would explain the male bias in the population. This finding is of great relevance to decisions about sustainable management. If there are many fewer females than males, then the prospect for sustainable use of the species is even more difficult. We have a species where females are larger and rarer than males. This is a nightmare scenario for management because hunters naturally would go for larger animals that give them more skin for trading. However, females represent both the productive sector of the population and the less abundant one! If I had any doubts about the practicality of this program, this result clinched it.

Now, if females indeed have higher mortality, I should be able to identify the reasons for it. Young anacondas suffer less predation as they grow larger. Just about any bird of prey or even many wading birds can prey upon a neonate-sized anaconda, but by the time they reach 1.5 meters (5 feet) their list of threats is reduced considerably. In my small study of neonates with transmitters, two of the five animals were preyed upon, and it so happened that both were females.[13] This is not a conclusive result given the small sample size, but it does point in the same direction of the model. Predation on medium-sized anacondas, 3 meters (10 feet) long, seems to be quite an important problem for males in the mating season since they travel long distances and move quite often during this time of the year looking for breeding females.[14]

Larger females are relatively immune to predation, though a large caiman might rarely take a weakened female, as in Olivia's case. So, we should expect less mortality in females than in males. However, the large expenditure in reproduction females incur might result in death not only as a result of predation but also during or after feeding due to weakening. Emaciated postbirth anacondas may fall victim to parasites and diseases more often than males. Furthermore, the environment itself can be very unforgiving: The largest animals may have to worry more about droughts and overheating.[15] As large anacondas incur higher locomotion expenses, they may be less able than males to track the receding waters in the dry season (Figure 4.6). Anacondas may hide under the mud even after it dries out and may survive weeks under the caked bottom of the dry ponds until the dry season breaks (Figure 4.7). But if a snake should venture out from the mud and try to reach another body of water, it must laboriously and slowly drag its large body across broad expanses of dry land, with no cover. Clearly, a snake with a larger body is slower on dry land. Males might be able to make it from pond to pond as the drought progresses due to their smaller size. In the treeless savannas of the llanos, this can be a dire undertaking as temperature can exceed 30°C (86°F) as early as 7:30 a.m. So, a snake might try an overnight trip, but if morning surprised it, overheating would be a serious danger. Unlike mammals and birds, reptiles do not sweat or have other ways to cool their bodies. The only air-conditioning option for a snake is to seek cooler places.

Therefore, despite my original resistance to accept the results of the model, these considerations allowed me to warm up to them. Perhaps the larger size of the anacondas and our thorough search for them allowed us a significantly higher encounter rate with females than with males, and that explains our bias in the raw data. In fact, the results of the model say that the encounter rate with females is higher than with males! So, I found fully internally consistent results that led me to accept them more favorably.

Anacondas' Health and Tribulations

One of my guiding principles for studying anacondas is to learn about them not from the human perspective but from the animals' point of view, to the extent that it is possible. This perspective helps remove the anthropocentric bias inherent in our species and gives us a better understanding of the animal itself. It's a little bit like young King Arthur in T. H. White's *The Once and Future King*: The magician Merlin turned him into other organisms so he could learn from them. Not having the benefits of Merlin's magic, all that I can do is a mental exercise and try to imagine what it is like to "be" an anaconda. However, as always, one needs to be careful what one wishes for. The life of anacondas is not an easy one even after the animal has outgrown most predators. My observations of large females revealed that diseases and parasites were routinely a burden. How fragile is the health of the anacondas? I wondered. How sensitive are they to the constant attacks of their little enemies?

The first aspect is the number of parasites they need to deal with. I sometimes found leeches stuck to the snakes, often inside recent wounds, or clinging to old wounds where scales no longer covered the soft skin. Anacondas found on dry land commonly had a heavy tick load (*Amblyomma dissimile*). Blood analysis conducted by my collaborator Paul Calle, a veterinarian, revealed that all 24 snakes examined had some sort of reptilian malaria (*Hemoproteus* sp.). That's a 100% infestation rate! Autopsy of two anacondas found dead revealed abundant cestodes, including *Crepidobothrium* sp., and trematodes (parasitic worms) in the jejunum and the duodenum in one of the specimens. Several apparently healthy anacondas also had subcutaneous nematodes (*Draunculus* sp.).[16] Do I really want to know what it is like to be an anaconda?

The amazing thing is that most of these animals looked healthy despite the diverse parasitic infestations they suffered. Some of them reproduced during that year, and others were recaptured, healthy and thriving, 4 and 5 years after the parasite infestation was first diagnosed. What is their secret to avoid being sick despite parasitic infestation? In theory, organisms reach a reciprocal co-evolutionary balance where parasites become less harmful and the host develops ways to accept them and live with them. Yet, much more research is needed in this field.

In some cases, I found dead animals but could not discern a clear cause of death (Figure 4.8). During autopsies, I noticed dark spots (~1 mm^2 [0.06 inches]) in fat tissue and elsewhere. The source of those spots was suggested by the fate of Sue, a 420-centimeter (total length [TL]), 44.5-kilogram female. I found her in April 1997 alive but not moving. Later that day she died. Other than having some small, old wounds on her tail that looked like capybara bites, she appeared to be a healthy snake. But histopathological analysis of her fat tissues revealed a type of lymphatic cancer (lymphosarcoma) that probably caused her death. I had caught Sue each year since 1992, when she was one of the snakes diagnosed with reptilian malaria. She had reproduced twice in the intervening years, and the abundant fat and early-stage eggs in her abdomen indicated that she was due to reproduce again that year. So, the cancer that seems to have claimed her life apparently did not make her sick!

Finding an anaconda with lymphatic cancer made me wonder a lot about causes. Lymphatic cancers in mammals can be triggered by the presence of chemicals and environmental pollutants. The llano is a flat savanna that receives no runoff from other lands, and there is hardly any agricultural activity, not in these floodable plains at least. In the study area there is definitely no use of insecticides or other agrochemicals. However, whistling ducks (*Dendrocygna* spp.) feed on rice fields in certain areas of the llanos where rice farming takes place. They may travel long distances tracking the receding waters as the dry season progresses. Might the ducks bathing in the insecticide-filled waters of the rice fields bring the pollutant on their feet? Or have they incorporated the pollutant into their tissues so anacondas might acquire it when they consume ducks in their regular diet? I have not found any other snakes with lymphatic cancer, so I put it on the back burner, although it leaps forward now and again.

A constant source of amazement and admiration was the capacity of the anacondas to endure and even thrive with major injuries. It is not uncommon to find an anaconda with an open wound bearing all kinds of things that should not be in a wound. Despite harboring leeches and copious amounts of mud these wounds did not look infected, were not swollen, and did not contain pus. In most cases wounded anacondas did not look ill or weakened; on several occasions, I ended up finding the animal again in a later year with the healed wound or even with the wound still open but without seeming to affect the animal's well-being.

An impressive example of this resilience was presented by María Eugenia (356 centimeters [12 feet] TL, 20.5 kilograms [45 lbs]). When I found her in March 1992 she had a pronounced ocular abscess (she did have one abscess with pus) from a twig that had penetrated the roof of her mouth (Figure 4.9). Other than the deplorable state of her mouth and eye, María Eugenia looked healthy and in breeding condition. I wired her with a telemetry transmitter since it was the beginning of the project and I did not know if I was going to find the 12 anacondas I was supposed to wire for the telemetry study. I figured that if she died, I could document the death; if she didn't, it would be an even more interesting piece of information. Prior to the surgery Paul Calle removed the twig and drained the abscess, to the extent it was possible, hoping to improve her odds of survival. The abscess was very large and seemed to have been there for quite a while. Part of the surgical procedure called for a preventive dose of broad-spectrum antibiotic (enrofloxacin) that we figured could help her plight as well.

I let go of María Eugenia wondering if she would be alive next time I found her. During the follow-up, I always tried to observe how the eye was doing, but it wasn't easy because she was often hidden underwater or in a cave. I did not want to overly perturb her or compromise the data on mobility and habitat use I was collecting. About 3 weeks after the surgery I managed to get a good look at her eye and confirmed that the swelling, and likely the abscess, was back to full size. Nevertheless, she survived, and thanks to the transmitter I could follow her throughout the year.

During the following months, I saw María Eugenia in a bank cave with a male with whom she might have mated. She did not move much at all for the reminder of the year. I didn't know it then, but such sedentary habits are typical for breeding females. They find a place where they can safely bask, and they pretty much stay put throughout their pregnancy. I did not see her give birth, but later in the following dry season I saw her very skinny. I thought her mouth injury had prevented her from feeding while she was pregnant, but later I learned that breeding females do not feed while they are gestating, so they often end up emaciated after giving birth. A month or two later she was spotted again with an inflated bulge in her midbody, evidence of a big meal. This put me at ease because I was reassured that she had recovered from her mouth injury and was going to survive. About 2 months later María Eugenia had another meal, and it was time to catch and measure her again before the radio died. You can imagine my surprise when I saw that María Eugenia had basically the same eye abscess she had when I first caught her, with a similar amount of swelling, but otherwise she looked very well indeed!

The case of María Eugenia was not an isolated one. In April 1995 I found a large male, Flemosito (3 meters [10 feet] long, 8.9 kilograms [19.6 lbs]) with a throat infection that had produced some phlegm and visible swelling in the back of the mouth. He wheezed when he breathed. He did not look very ill, his body was in good condition, and I thought it was an affliction he could easily shake. However, Flemosito was found again in March 1996 in a similar location and, sure enough, he was still wheezing away. He had not grown any in length and had lost a little mass (450 grams [1 lb]). Flemosito had been caught in 1993 for the first time and did not grow at all in 3 years, which is common in males of this size. Two months later we found Flemosito's tail floating in the area; I recognized it by the pattern of subcaudal spots I recorded for each snake. He had apparently been eaten by a caiman. He had survived the small enemies but succumbed to a large one.

So, anacondas seem to be able to survive parasites and diseases that would render a human bedridden, miserable, and unable to function. Did anacondas feel comparably sick? Does it bother them as much as it bothers us to be sick? There are clear limits to how much I can learn about being an anaconda. I have on occasion mused that it was possible to learn everything there is in biology through the detailed study of the life of one's beloved animal. I don't know if this is possible, but I will be pushing the limits of my own longevity to fully understand the field ecology of anacondas, long-lived as they are. How wonderful would it be if we could break the secrets of their immune system and learn how their mysterious physiology fights and endures diseases and wounds!

The Elusive Babies

As the project progressed, I spent day after day trudging through the swamps and finding anacondas in their native habitats. The radiotelemetry observations gave me more and more insight into what the life of anacondas was like, and I felt I was really making progress in disentangling the secrets of this fascinating snake. However, there was a part of the puzzle painfully missing: Where were the babies? Of course, my method of looking for animals by feeling for them under the vegetation or seeing their shape under the water or sticking out of the mud was bound to overlook the really small ones. I already knew this to be a problem in detecting males, and with babies, which are so much smaller, this problem was expected to be a lot worse. Once I found one and had one of my collaborators hold it under the vegetation so I could step on it gently to see if I could notice any difference between how it felt to step on a baby versus a stem of water hyacinth. There was no difference at all. Their blending in with the environment is visual and also tactile! Then I realized that the apparent lack of babies in my data could simply reflect the difficulty of detection.

Studying babies is a problem for many herpetologists. Baby reptiles and baby snakes in particular have been very difficult to study in the wild precisely because it's so hard to find them in the wild. But, at more than 200 grams (0.5 lb), baby anacondas were as large as most natricine snakes and three times the size of baby pythons. Anacondas represent perhaps the best bet at learning about neonate snakes we are going to have.

What is a neonate?

Defining a neonate is straightforward in most other animals but not in anacondas. Age after birth is a criterion for neonates commonly used in most species. A neonate's age has been defined as 10% of the time it takes the animal to reach adulthood.[17] To use this definition, of course, we need to know how long it takes the animal to reach adulthood, so it didn't help much for my attempts to study or even define a neonate. To further confuse this definition, anacondas have a very strong sexual size dimorphism,[15,18,19] with females being much larger than males (see Figure 4.2), so females may take a lot longer to reach adult size than males. Thus, a neonatal female could be far larger than a non-neonatal male!

In view of these difficulties, I adopted a size-based definition of neonate, encompassing neonate and neonate-sized individuals regardless of their age. An older neonate-sized individual is an animal that may be facing the same ecological scenario and challenges that a true neonate does in terms of temperature regulation, predators, and prey availability. However, it differs from a neonate *sensu stricto* in that an older individual has had time to learn, and its developmental and cognitive abilities may be somewhat different than a neonate's.[20] Size, then, is not a trouble-free criterion for defining a neonatal anaconda. There is great variability in the size of neonates at birth depending on the mother's size: Larger females bear neonates that exceed 90 centimeters (3 feet) in length, while the neonates from smaller females may be only 65 centimeters (25.5 inches) TL or less. Some neonates can weigh as little as 150 grams (5.3 ounces) at birth, while others can weigh as much as 330 grams (11.6 ounces), a twofold difference in size within the neonate class.

Over the years of working with this species I had the strong impression that animals smaller than 1 meter (3.3 feet) snout–vent length (SVL) were in a class of their own and ecologically different from the rest of the population. Not wanting to set an arbitrary limit, though, I plotted the cumulative frequency of the different sizes from the 788 animals caught over 17 years of study in different locations. Looking at the distribution of data, it seems as if 92 centimeters is a natural cutoff.[13] Larger animals are almost certainly older than 1 year, but smaller animals are not necessarily younger than 1 year.

Because of the large size of neonatal anacondas, I had the unprecedented opportunity to wire neonates and follow them with radiotelemetry. I radiotagged five neonates I obtained from females that gave birth in captivity and followed them for about a month. This was not my original intent: The transmitters were supposed to last 4 months, but all of them failed after about a month (they were made by a company that produced transmitters that failed in four other projects with reptiles that same year). Thus, a lot of my learning with the neonates comes from the opportunistic capture of the wild population, some observations done in naturalistic enclosures,.

No Hyperkinetic Kids: Behavior and Mobility of Baby Anacondas

During one of the years of my study I held two pregnant female anacondas in captivity until they gave birth and had the opportunity to radiotag some of the neonates. Clearly, I had to use a smaller transmitter for these, which resulted in a shorter life span of the battery, but it gave me the opportunity to study neonates, which until then had not been done with any snake. It wasn't difficult to keep up with these baby anacondas: They moved very little, consistent with observations in captivity. About 90% of the activity registered by the neonates was "inactive," often hidden in caves or under the aquatic vegetation. Sometimes they showed short bouts of activity, toward the evening or when a temperature drop might suggest nightfall. When a neonate is in the water it usually has its head on the water edge facing the land in what appears to be a hunting attitude. Neonates spent most of their time in the water, and 78% of the time the animals were in the water they had their head on the border, which is a typical sit-and-wait position. This brings back observer bias when we study an animal that is very different than ourselves and don't really know what it may be feeling, sensing, or even thinking. Because of their sit-and-wait hunting strategies, it is quite possible that many times when one sees a snake that is "inactive" or "immersed," the snake could simply be "hunting."

Wild neonates were often seen moving on the road, catching prey, or basking. This is fairly different from the reported activity of adult animals but is probably a consequence of differential detection. Clearly, neonatal anacondas that are moving on the road, killing prey, or basking are more exposed and have a higher likelihood of being detected than inactive (whatever that means) ones. Babies have no venom, no speed, and no big fangs, and they are not big, yet. They rely on blending in to survive predation, so immobility is the best way to avoid detection. A neonatal anaconda is pretty much like a chocolate cookie: Whoever grabs it, eats it. So, all the neonates that I found were lucky that it was me who found them, not some predator.

Neonatal anacondas were found to use stagnant, shallow (less than 20 centimeters [8 inches]) water under plant coverage. This is the same kind of habitat that adult anacondas regularly use.[4] The only instances in which radiotagged individuals were found anywhere other than the place where I had released them were with Cassie and Ingrid. Cassie was first observed in a small creek that connects the borrow pits with a main river, 360 meters (394 yards) east of the former location. The next time I looked up Cassie, the radio signal led me away toward the dry land, to a tree that held an active nest of crested caracara (*Polyborus plancus*). I found the chewed-up transmitter at its base.[13,14] Cassie was one of the neonates that showed a great deal of movement during the time I followed her. The other individual that showed higher mobility, Ingrid, was also preyed upon. It is likely that their high mobility increased their chances of being detected by predators. During the telemetry study I also found a tegu (*Tupinambis teguixin*) eating a neonatal anaconda, but I didn't see the capture, so I don't know if the tegu found the dead neonate and was just scavenging or if it actually preyed on a live baby.

Compared with older conspecifics, neonatal anacondas have a much smaller home range. This is expected as the larger individuals need a larger home range to find all the food they require to meet their higher energetic needs. Also, after they achieve a certain size, the number of predators that threaten them is much smaller and they can move more freely through different areas. Conventional wisdom suggests that juveniles spend more time and energy foraging to reach adult size sooner, but doing so increases their risk of predation. This is probably what happened to Cassie and Ingrid. This may be the reason that the few babies that were recaught (and survived) had not grown much, perhaps associated with very low mobility.[13]

The difference between the mobility of Ingrid and Cassie and the other neonates was striking. Their high mobility likely led to their being preyed upon. Some individuals might forage more actively, while others might spend more time in safer places, forfeiting potential meals and growth. Animals that forage more actively might be better adapted for habitats with few predators or different conditions. Females might hedge their bets in these unpredictable habitats by producing animals with different strategies. If the conditions are favorable for animals with more activity (e.g., fewer predators), active animals might be more likely to survive and reach breeding size earlier. If, on the other hand, predation pressure is high, then active animals would be become prey more often, with a resulting selection for animals that forage less actively (see Chapter 6).

Get Tough Soon or Else!

Baby snakes are difficult to find. Of a sample of 696 individuals of the core study site, only 28 (3.59%) fell within my definition of neonate. Of these 28 animals, only 2 (7%) were recaptured in a later year. In comparison, the average recapture rate from year to year in the adult population was 24%. This low recapture rate of neonates supported the idea that neonates are just not easy to find. Of course, the lower recapture rate of neonates could also be associated with a lower survival probability of this size class. This latter hypothesis is supported by the fact that I did not catch these neonates, at any size, after several years. Clearly, if the problem was only low capture rate for their small size, I would have found some animals later after they had grown up. Other than the recapture of wild animals, I also marked 278 neonates born in captivity from wild-caught females. None of these neonates were found in any of the following years. This strongly supports the notion of very high mortality for baby anacondas.

Conventional wisdom says that neonates need to grow quickly, which helps their survival, until they reach reproductive size. However, the only two neonates that I recaught in a later time had a very low growth rate in general terms. The average growth rate was 0.071 millimeters/day (0.03 inches/day) in an 11-month period.

The growth rate found among neonates appears low for animals that are expected to be in their peak growing period. This growth rate is also low when compared to the growth rate of captive neonatal anacondas fed regularly. I estimated growth rate from three neonates raised at the San Diego Zoo using their mass when they were acquired to estimate their length (length data were not available). Growth in the first 504 days is estimated to be 0.14 millimeters/day (0.006 inches/day), twice that found in wild animals. Other captive anacondas show a neonatal growth rate as high as 2.13 millimeters/day (0.08 inches/day) in the first 445 days.[13] Although wild animals have a large availability of natural prey, foraging may result in higher predator exposure. It is likely that the neonate's food intake is limited to the few animals that move within their reach in their protected hideouts. More adventurous neonates that actively seek food may become prey themselves. This high predation would also account for the low recapture rate for neonates. The strong seasonality of the llanos might select for foragers that do not range widely but wait for their meals in a safe place. Animals might forfeit growth in exchange for a higher survival rate. This strategy would be beneficial in places with a strong dry season because neonates that venture out of their hideouts would be more exposed to predation. On the other hand, the same drought would make water bodies an oasis where all small birds and potential prey come for water (Figure 4.10).

One of the wild recaptured animals, Tuertico, was missing one eye at the time of the first capture. Surprisingly, he was still alive the next year despite this apparent handicap. Tuertico's daily growth rate was 0.068 centimeters/day (0.027 inches/day), not much lower than that of the other individuals (0.074 centimeters/day [0.029 inches/day]). It is surprising that the missing eye did not seem to have a strong influence on his growth rate, or his survival for that matter. Tuertico was found in the second year in the same cave where he was found the prior year. His low mobility might be associated with both his low growth rate and his survival despite missing one eye.

Mom, when's supper?

Eight of the 28 (28.57%) neonates found were either consuming or had recently ingested a meal. This percentage might not accurately reflect their feeding frequency since animals that are constricting prey or have had a recent meal are more conspicuous. Neonatal anacondas feed on relatively large prey (Table 4.1); the average prey size is 26.3% of the snake's premeal mass. This is not a surprising proportion compared to adult anacondas,[21,22] and it is certainly not uncommon for large constrictors.[15,23-25] A 25% relative prey mass is common in other neonatal water snakes (Nerodia sipedon).[26] However, I believe that because they are ambush hunters, they cannot be too choosy in their prey preferences and might have to take whatever prey becomes available. Neonatal anacondas can take a prey as large as 42% of their body weight. This in itself may be a risk for the babies that endure long periods of hunger and then try to attack what is a very large prey for their size. Either the prey, while defending itself, can hurt the neonate or the neonate might become vulnerable to a predator after taking too big a meal that makes it more conspicuous and hinders its ability to escape.

Getting Along

Many people are surprised to learn that several anacondas can often be found in the same pond or cave, getting along rather well. The conventional wisdom suggests that, being predators, they should avoid each other to lower competition for potential prey. Furthermore, since they can on occasion practice cannibalism, the question of how well they get along is not a trivial one.

That anacondas tolerate each other's presence very well was known even by the early explorers of the Amazon. Lange,[27] in his early account of exploring the Amazon, writes:

Table 4.1 Neonatal anacondas in the Venezuelan llanos found with evidence of a recent meal.

ID Number	SVL (cm)	Mass (g)	Prey item	Prey size (g)	Relative prey size (%)
Mammals					
E247C	73	262.25*	Unknown mammal	112.75	42.99
E498C	81.5	363.19*	Unknown mammal	96.81	26.66
Birds					
E1008C	58.03	180	*Jacana jacana*	40	35.28
E1009C	82	235	*Crotophaga ani*	100	43.5
E1012C	75.43	440	*Jacana jacana*	80	18.18
E738C	73.8	315	*Crotophaga ani*	100	31.75
E861C	76.5	301.19*	*Passerinae*	68.81	21.8
E588C	81.8	460	*Jacana jacana*	70	15.20
JL-24	73	195	*Phacellodomus rufifrons*	21.5	11.03

When the prey mass was unknown, we estimated the mass of the prey by subtracting the estimated mass of the anaconda based on its SVL. These cases are indicated with an asterisk (*). In this table we also include a neonate that was born in a naturalistic enclosure (JL-24) and that caught a bird before it was found in the birth cage.

Reproduced from reference 13 with permission

"There we behold a writhing mass of long, blue-grey snakes of all sizes and lengths, some no more than six or seven feet, others over three times that length. At first they do not notice us. They are partly sleeping, some with their bodies under the surfaces of the lukewarm shallow ooze, while other mounted on the bodies of their brother snakes, or half creeping and resting on the margin of the pond, are happy, so to speak, in the family bosom. Some of them are as thick as a man's leg, while others are thin and slender as eels. The whole mass of this snake nest, however is at rest. A few of the smaller ones are lazily creeping in and out through the labyrinth of snake bodies." (p. 134)

This report probably refers to a breeding aggregation, one of the most interesting aspects of anaconda biology (see Chapter 8). But, unfortunately, it ends with the explorers taking aim with their weapons and killing just about every snake they could, as was the practice in the early days of the exploration of the tropics.

In a much less unfortunate encounter for the snakes, on one occasion I found 34 juvenile anacondas (less than 4 kilograms [9 lbs]) in a 400-square-meter area, the size of a regular school classroom. Due to the rarity of finding young individuals and the large number of them in such a small space, I called this "the kindergarten." I captured all the animals for processing but returned them back and unhurt when I had finished collecting data. I am in the process to figuring out why these animals were together by looking at their level of genetic relatedness. Were they related by blood, or was it just a random assortment of unrelated snakes that happened to be in the same place at the same time? They were larger than neonates, and likely not from the same clutch. Were they cousins or extended family? There might have been a social facilitation, meaning that they found each other and chose to hang out together. It is also possible that they just gathered in the last remnant of water of the creek and happened to be in the same place.

Conventional wisdom says that neonatal reptiles are miniature replicas of the adult individuals, and in the case of green anacondas this claim seems to be well supported. Neonatal anacondas are similar to adult individuals in their habitat use and preferences, mobility, prey size and prey preferences, time budget, sociality, and even growth rate. The main difference between neonates and adults seems to be that adults are less worried about predation.

Coming of Age: Growing Up to Be Different

The issue of growth is an important one with anacondas. On the one hand is the obvious case of how large they grow, which has earned the anacondas their mythical place in folklore and popular culture. But looking into it in more detail, we see that anacondas are born weighing around 200 grams (7 ounces) and can grow up to weigh more than 100 kilograms (220 lbs). This represents a 500-fold increase in body mass, the largest in any snake species. This constitutes a tremendous ecological change in the individual not only regarding predators and prey but also thermoregulation and habitat use. Based on the few data I have, neonates, the class that is expected to grow the most, show an extremely low growth rate. This low growth could be associated with the low mobility and low foraging. What is the growth rate like in older individuals? Is it possible that as they get larger they become able to forage more actively and enjoy a higher growth rate? Is there a size where the growth rate exceeds that of neonates? For how long do they grow? Do they stop growing at a certain age or after reaching a certain size? What size do they really reach? How long does it take for them to reach those sizes? Not having answers for most of these questions, I obviously have barely scratched the surface.

The data on growth are difficult to interpret because the trends are not clear. Smaller animals, larger than neonates, seem to grow faster than older ones, but the trend is not clear at all, and there seems to be a lot of individual variation, including a large variance in growth in the same size range. Thus, it seems as if an animal's growth rate does not have a lot to do with its size. There are many animals of all sizes with a growth rate of zero or near zero. There are animals with low as well as high growth rates in just about every size class, which suggests that there may be two different strategies that may be independent of the size of the animal. Clearly, if some animals are meant (as in genetically prone) to grow quickly and others are meant to grow slowly, there would not be much of a relationship between size and growth when we consider the whole population.

One thing that strikes you when looking at the anaconda population is the sexual size dimorphism found in the population: Females get a lot bigger than males. Basically, males pretty much stop growing, or grow very slowly, when they reach an SVL of 244 centimeters (about 8 feet) and a weight of 7 kilograms (15 lbs), while females continue growing much larger. Very few females breed at less than 275 centimeters (9 feet) TL, so there is hardly any overlap in size among sexes in adult anacondas. In fact, it is possible to identify the sex of an adult with great confidence only by looking at its size (see Chapter 8).

This sexual size dimorphism is not present from birth, however: There is no difference between the size of boys and girls at birth. When we look at the population as a whole, males grow much more slowly (0.018 centimeters/day [0.007 inches/day]) than females (0.036 centimeters/day [0.014 inches/day]) in terms of SVL. What is producing this growth difference among the animals? Very young males grow as much as 0.07 centimeters/day (0.03 inches/day) when they are neonates, but their growth is reduced significantly shortly afterward. Males seem to reach an optimal size that increases their reproductive success, and once they reach it there is not any noticeable growth (see Chapter 7). Many males show no growth or grow very little after reaching a certain size.

The apparent stoppage in growth occurs at very different sizes in males and females and seems to be the reason behind the dimorphism found in the species. Some females seem to stop growing at 3.5 meters (11.5 feet), while others grow past that size in relatively few years, supporting the notion that there may be different growth strategies in the population. However, the stoppage in growth in some individuals may be only apparent. For instance, the female Guaratarita was caught in 1994 at 362 centimeters (12 feet) TL. She was caught again in 1996 and had not grown at all. Then she was caught again in 2007 and measured 407 centimeters (13.5 feet). Guaratarita grew only 40.5 centimeters (1.3 feet) in 11 years (0.011 centimeters/day [0.004 inches/day]). This is quite a slow growth rate for an animal that is in a "growing phase" and that would obtain tremendous benefits from being larger, as fecundity is correlated with size (see Chapter 6). Since I did not catch Guaratarita for 11 years, I don't know if she simply grew slowly during all these years or if she stopped growing for a few years and then had a short bout of intensive growth during a few of the years in which she wasn't found.

Reptiles are supposed to grow throughout their life, although their growth rate is expected to slow down later in life. My data support this notion. Many adult animals recaught at adult size for several years showed little to no growth in as many as 5 years. While most males stopped growing at a determined size, this is not the case in many females. Many females show relatively steady growth rate, while others seem to stop growing. No animal was found to be much larger than 5 meters (16.4 feet) long, and there seem to be ecological limits preventing females from growing larger than this size (see Chapter 6).

How Old Do They Get?

Perhaps one of the most common questions people ask me about anacondas is how long they live. I don't have a definitive answer as they seem to live longer than I have studied them for, but knowing their growth rate and the size they reach, I'm prepared to make an educated guess. A male anaconda that is born at, say, 68.42 centimeters (2.2 feet) SVL would grow at 0.07 centimeters/day (0.03 inches/day; neonatal growth rate). Let's assume that he grows at this rate until the size that he starts reproduction, 159.3 centimeters (5.2 feet) SVL. Thus, it would take approximately 3.5 years for a male to reach reproductive size. Then, if we use the average growth rate of adult males (0.018 centimeters/day [0.007 inches/day]) to calculate growth thereafter, we find that it would take 9.9 more years for a male to reach the average male size (225.9 centimeters [7.4 feet]), and it could take as long as 12.12 years to reach the maximum SVL (307 centimeters [10 feet]) I have found in males, for a total of 25.6 years.

On the female side we find a similar scenario. Using a rate of 0.07 centimeters/day (0.03 inches/day) to calculate growth until she reaches reproductive size (210.7 centimeters [7 feet]) and the average growth rate for females (0.036 centimeters/day [0.014 inches/day]) thereafter, we find that a female would need 5.5 years to reach sexual maturity, 8.8 more years to reach the average female size, and 11.6 more years to reach the maximum recorded size, for a total of 26 years. Using regression models obtained from the growth equations, we obtain similar figures for first reproduction and average size but much higher ages for maximum size with both sexes.

However, these calculations use rough averages, and there might be significant differences in particular individuals. In the case of Guaratarita, if we use the average growth rate of neonates until first reproduction and the documented growth rate I found in her since, it suggests that she could have been 36 years old when I first caught her in 1994 and 50 years old in 2007. At this growth rate, Guaratarita would need to be 73 years old if she were to reach the size of 5 meters (16.4 feet) long found in some animals of the area!

Does an anaconda really need 70-plus years to reach the 5-meter (16.4-feet) mark? How old would an anaconda have to be to reach, say, 7 or 8 meters (23 or 26 feet)? The quote with which I started Chapter 2 from a former professor at Universidad Central de Venezuela, Dr. Francisco Mago-Leccia, an ichthyologist, stressed the need for long-term, up-close and personal interaction with the animals to really learn about them. I have been studying anacondas for a bit more than 25 years, so I am ill prepared to make guesses about 50 years of growth the animals may experience. As some of the evidence suggests, animals might use different ecological strategies. Some seem to grow relatively quickly, stop for a while, and then resume growth, as might have been the case with Guaratarita, but it is possible that other animals grow at a higher rate earlier in life and then stop growing altogether after reaching a certain size. As suggested by Olivia's only recorded size of 5 meters (16.4 feet), there might be a size limit of about 5 meters (16.4 feet) that anacondas cannot pass in the llanos (see Chapter 6). Of course, it is possible too that they simply grow slowly all their life and the really large animals that have been reported out there are Methuselahs that have lived for many decades. This is another reason to revere and respect these majestic giants.

5

How Much Is Supper? Predator Diet and Prey Retaliation

No obstante el profundo ensimismamiento en que iba sumida, doña Barbara refrenó de pronto la bestia: Una res joven se debatía bramando al borde del tremedal, apresada del belfo por una culebra de agua cuya cabeza apenas sobresalía del pantano.

Rígidos los remos temblorosos, hundidas las pezuñas en la blanda tierra de la ribera, contraído el cuello por el esfuerzo desesperado, blancos de terror los ojos, el animal cautivo agotaba su vigor contra la formidable contracción del los anillos de la serpiente y se bañaba de sudor mortal.

"Ya ésa nos esa escapa," murmuró doña Barbara. "Hoy come el tremedal."

Por fin la culebra comenzó a distenderse sacando el robusto cuerpo fuera del agua, y la novilla empezó a retroceder batallando por desprendérsela del belfo; pero luego aquella volvió a contraerse lentamente y la victima extenuada, cedió y se dejó arrastrar y empezó a hundirse en el tremedal lanzando horribles bramidos y desapareció dentro del agua pútrida, que se cerró sobre ella con un chasquido de lengua golosa.

(Despite the deep self-absorption in which she was immersed, Doña Barbara suddenly held her horse: a young heifer struggled bellowing at the edge of the quagmire, imprisoned by its lip held by an anaconda whose head barely stuck out of the mud.

With stiff and trembling limbs, its sunk the hooves in the soft ground at the edge of the swamp. Its neck tightened due to the desperate effort, white with terror the eyes. The captive animal exhausted its strength against the formidable contractions of the coils of the snake and drenched itself in deathly sweat.

"There is no escape," whispered Doña Barbara. "The quagmire will feast today."

At last the snake started to stretch, sticking its robust body out of the water, and the heifer started to back up, battling to free its lip; but then the snake contracted again slowly and the victim, already exhausted, gave up and let herself be dragged and started to sink in the quagmire, emitting horrible bellows, and disappeared in the putrid water, which closed over them with a snap of its glutton tongue.)

Excerpt from *Doña Barbara*[9] pp. 466–467 (Translated by JAR)

I will never forget the countless hours I spent as a child at Parque del Este, one of Caracas's main zoos, where I eventually was lucky enough to apprentice as a zookeeper, watching anacondas in the big containment pit where they were housed. I would strain my eyes trying to identify the massive snakes under the murky water. Even though the snakes did not move and were acting very much like logs, my 7-year-old imagination was too excited to be bored by the static view. I was imagining the body of the corpulent serpent under the water and wondering how it would kill a large prey. Could it kill a jaguar? A tapir? A crocodile? I knew capybaras and otters were well within their range. Among the many questions that I mercilessly shot at my mother, a psychologist, about anaconda biology, many were, obviously, about eating. She did her best to satisfy my curiosity with the knowledge she had and told me what any non-biologist could have said.

I still remember something she told me that I later heard many times from the llaneros in my study of anacondas: "The anaconda kills a white-tailed deer, but being unable to swallow the antlers she would swallow it butt-first and work her way all the way to the neck. Once there, the snake would wait with the antlers sticking out of her mouth for 3 or 4 months until the neck rotted and the antlers fell off. Then the snake would finish swallowing the deer." This was so fascinating that I could easily envision the snake's body engorged by the recently swallowed deer, and I could picture vividly the dead deer with its open eyes sticking out of the snake's mouth. I borrowed the mental image from some deer I had seen in a diorama at the museum of natural history and mentally pasted that head onto the mouth of the snake that I imagined lay in ambush under the murky waters of the exhibit. It was as good as if I had actually seen the snake with the dead deer! Years later, through my research, I had the opportunity to actually test this hypothesis experimentally.

Of course, the academic interest that one might have about the anaconda's diet pales by comparison to the interest that the anaconda awakens as a potential predator for people. Many of the people who regularly contact me with questions are concerned about whether anacondas can eat people. Clearly, there is something very personal about dealing with an animal that has the potential to feed on one's flesh, and this often inspires feelings of awe as well as fear. These feelings can get in the way of understanding the biology and the feeding ecology of the species. In this chapter I will share what I have learned about the anaconda's diet and what role it plays in the big picture of the anaconda's biology. Unable to avoid the admiration I have for these serpents, I will try to use it to my advantage in understanding the feeding ecology of the anaconda and identifying the aspects that really matter for the snake.

Where and How to Get a Meal

Clearly, food intake is a critical aspect of any animal's ecology, for it determines how the animal obtains the energy needed for survival and reproduction. How the animal goes about obtaining this energy is critical to the species ecology. Anacondas, with their strong build, are not fast enough to pursue prey on land. Although they are fast swimmers, an open mouth moving forward is not very hydrodynamic for catching fish, which are often quite fast. However, anacondas can launch an extraordinarily fast attack on a prey out of the water when needed. In fact, the predatory strike of an anaconda can be so fast that you can't even see it happening. Once I was watching a small heron on the swamp with binoculars and, right in front of my eyes, the heron became a knotted bird in the coils of an anaconda (Figure 5.1). The strike was so fast that I didn't see the snake grabbing it, throwing its coils around the bird, and finally subduing it. My brain didn't register any of it. The heron went from walking to captured in a split second.

Anacondas' preferred strategy is to wait in a place under the vegetation, with which they blend wonderfully, and wait for the right prey to come within range. This way, anacondas avoid being detected moving around looking for prey and don't need to expend the energy needed for locomotion. As reptiles, anacondas also have a very slow metabolism, which means that they consume very little energy for metabolic maintenance. Thus, they can remain for quite a long time in one place waiting for the right prey to appear. This is the reason they cannot be easily caught in traps (however you might want to concoct a trap for anacondas): Because they don't move much, they are not likely to find the trap by chance, and because they don't actively seek their prey, they are not likely to be lured by bait.

Killing the Prey: Not Just a Big Hug

In my quest for knowledge from the llaneros, there were always some stories I had to filter out. In particular there was one story that I knew I'd hear whenever I asked a cowboy about anacondas. They would all tell me how anacondas can kill a big bull, not unlike the quote from Doña Barbara at the beginning of the chapter but far more dramatized. As the story goes, a 700 kilogram (1,540-lb) bull comes to the water and the anaconda, with its tail anchored to a log under the water, grabs the bull by the nose. The bull pulls back, trying to free itself, but the snake pulls him back in the water. As the tug-of-war continues, the snake stretches, becoming as thin as 2.5 centimeters (1 inch) across (down from 45.7 centimeters [18 inches] or so) and can stretch as much as 50 meters (54.7 yards), only to pull the bull back to the water again. This struggle continues until the bull is exhausted and surrenders to the snake. The snake then wraps itself around the bull, squeezing and grinding its bones and rendering it shapeless. The sound of the bones being ground up is quite chilling. Once the bones have been ground up, the anaconda consumes its shapeless prey, which now can be easily swallowed.

This was a good example of the many tall tales I heard about anacondas that were not true, and it was likely the story that Rómulo Gallegos heard that inspired the excerpt I quoted at the beginning of the chapter. The reason I refer to it here is because it is so common: Never once did I fail to hear this story when I asked anyone about the anaconda's diet. Yet I never found a person who could provide a firsthand account. People always said they'd heard the story from someone who was "very trustworthy." Of course, like many other myths, there are bits and pieces of truth to it. Anacondas do anchor themselves on logs in the bottom of river to pull prey into the water. Anybody who has skinned a dead snake, which llaneros likely have, knows how elastic its skin is; compared with other animals, but not quite as extreme as in the story, it is the skin that stretches; the snake's body as such does not stretch nearly as much. Llaneros are obviously familiar with how sensitive bulls' noses are, and they know that even a large bull can be dominated by holding it by the nose. Of course, llaneros do not know that between a bull and a snake, the snake will become exhausted a lot sooner than the bull, since a mammal's aerobic metabolism is much more efficient than that of a reptile's in terms of how quickly they process energy. Also, the constriction of the snake does not grind the bones, nor can it accomplish such a feat. However, when a snake is killing a prey, it is possible to hear cracking, popping noises as some ribs may break and the joints and articulations are disjointed, part of the subduing strategy the snake uses.[28]

Other tall tales about anacondas killing their prey can be found in the writings of Padre Gumilla, a priest who came to the Americas in the 1700s and wrote interesting accounts of what he found.[1] pp. 376–377 As he narrates:

Es el caso que al sentir ruido levanta la cabeza y una o dos varas de cuerpo; y hace la puntería al tigre, león, ternera, venado u hombre; luego abre su terrible boca arroja, sin errar la puntería, un vaho tan ponzoñoso y eficaz que detiene, atonta y vuelve inmóvil al animal que inficcionó; le va atrayendo hasta que dentro de su boca a paso lento, si, pero indefectiblemente se lo traga si alguna casualidad no lo impide. Dije que traga porque no tiene dientes, y gasta largo tiempo, y aun días enteros, en engullir una presa; y es tal y tiene tales ensanches su fatal gaznate, que a fuerza de tiempo se traga una ternera de año, estrujándole la sangre, y el jugo al tiempo que la va engullendo.

(This is the case that when it feels a noise, it raises its head and one or two yards of its body, and aims at the tiger, lion, heifer, deer, or man; then opens its terrible mouth and throws without missing its aim, a whiff so poisonous and efficient that stops, stupefies, and paralyzes the animal that it targeted; it starts attracting it until it is inside its mouth, slowly, yes, but indefectibly* it swallows it unless some accident prevents it. I say it "swallows" because it has no teeth, and spends a long time, even whole days, in swallowing its prey, and it has such width on its deadly gullet, and using so much time, it swallows a one-year-old heifer, squeezing out its blood, and its juice as it swallows it.)
* unfailingly

(Translated by JAR)

Clearly this description is not accurate, but you'd be surprised how many people believe that anacondas, or boas, expel a hypnotizing breath that paralyzes the prey and helps the snake to consume it. It is uncertain if Padre Gumilla heard this story from the people or the story spread to the public from his writing. But let me give you the straight story now.

How Do Anacondas Kill Their Prey?

During their evolutionary history snakes lost claws and limbs. They modified their dentition, giving up the cutting teeth their ancestors, lizard-like animals, had. In the process snakes also became slimmer than their ancestors, likely an adaptation to their lives under the ground, where being slender allows animals to fit through small holes. The consequence of becoming thinner, with the same energetic requirements, is that they had to consume wider prey in comparison with their own diameter. To cope with this new challenge snakes developed a streptostylic jaw, which refers to the many adaptations involving mobile joints (compared to the immobile ones of other animals) that allow snakes to swallow large prey (see Figure 2.18). The jaw muscles in snakes are far more posterior than in their lizard relatives, which results in less efficient bite force (David Cundall, personal comunication). The snake's head is like a high-tech device, and like any high-tech device it is very specialized—it's very good for one thing but not good for others. Due to all these mobile joints needed to swallow large prey, snakes gave up the crushing power of lizard jaws, which leaves them with a new problem: how to kill their prey. Not having cutting teeth and not having a crushing bite, snakes need an alternative way to kill their prey.

The question of killing the prey is relatively simple when the prey is a smaller animal but is more formidable when we are talking about large and dangerous prey. Few people realize how dangerous it may be for a snake to take a large meal. When a snake catches a prey, certainly if it is a large prey, the snake may be at great danger. Once a prey is caught, the snake has a feisty animal that is struggling for its life. Some snakes avoid large prey altogether. Other snakes have evolved powerful venoms that allow them to do an instantaneous inoculation, let the prey go, and find it later after the venom has done its job. This is the case of pit vipers and other venomous snakes. Constrictors, on the other hand, need to hold on to their struggling prey and kill it using their own limbless, clawless bodies without the benefits of powerful jaws or cutting teeth. This is not an easy task if you think about it. This same body contains all the vital organs that most animals go to big trouble to protect.

This is also very different than the relationship we have with our food. When was the last time you were afraid that your pizza would hurt you before you ate it?

When an anaconda attacks a prey item, it wraps its powerful body around it to kill it by constriction. The killing of the prey occurs by whatever mechanism does it first. The one that kills most prey soonest is circulatory arrest. Since filling the heart is passive (blood trickles from the systemic circulation into a relaxed heart), the snake's constriction of the heart prevents it from filling with blood.[29] The heart continues beating, as it has an internal pacemaker, but not being able to expand when it relaxes, it doesn't fill up with blood and the cardiac output is reduced to near zero very quickly. So, the constriction produces nearly instant circulatory arrest. Of course, not all prey are as easy to kill. Some are very difficult to constrict. Strong muscular animals are more difficult to get to the position where the constriction can stop the heart. Other animals, such as turtles, simply cannot be killed this way. The easiest way an anaconda can kill a turtle is drowning it by holding it under the water. This, however, is easier said than done, considering that many of these turtles are aquatic turtles that can survive for quite a long time without breathing, possibly hours. Yet, anacondas live in a different timeframe than we do. The snake can simply hold the turtle underwater for as long as it takes, while it breathes with its longer body out of the water (Figure 5.2). Swallowing a prey may take 3 or 4 hours, so spending another few hours to kill it is not necessarily a long wait.

The constriction of the anaconda involves more than just killing the prey; it's also about subduing it to prevent it from escaping and hurting the snake in the process. On occasion the constriction can be so strong that it can break the ribs of the prey.[28] A broken rib can, of course, pierce the heart or lungs of the prey, either of which would expedite its death. I found this to be the case in a young white-tailed deer (*Odocoileus virginianus*) an anaconda had consumed. The snake regurgitated the prey after we caught it, and the exam of the regurgitated fawn showed that the fourth ribs had been broken (Figure 5.3). One basic part of the subduing behavior is bringing the prey into the water, which helps drown the prey. It also brings the prey to an environment where it might not have the same locomotion skills, as many of the anaconda's prey are terrestrial.

Another characteristic of the constriction of the prey that is seldom noticed is that the snake also pulls and twists the spine. It is possible to see in a prey that is being squeezed how it may become longer as the snake's coils pull and twist the spine. This seems to have been the case with a predation event reported on a white-collared peccary (*Tayassus tajacu*) when the authors mentioned the cracking sounds produced when the anaconda was killing the prey.[30] It may have been also the case with a baby capybara that I found after an anaconda had constricted it but failed to consume it.[28] That the spine might be broken in the constriction process seems to be apparent in Figure 5.4, which shows an anaconda constricting a spectacled caiman (*Caiman crocodilus*) more than 2 meters (7 feet) long. Clearly, caimans are difficult to kill by drowning, and their powerful jaws can inflict great damage on the snake. By breaking the caiman's spine, the snake can reduce its ability to fight back and to escape. Not that breaking the caiman's back that badly is not an extraordinary feat in and of itself!

Not into Fast Food

Something we really need to remember in order to understand the anaconda's mind is how little hurry they are in when feeding. In today's world we dedicate increasingly less time to feeding ourselves and keep thinking of tricks to save time in meal preparation, so it comes as a shock when we realize how much time anacondas take to eat their prey. A big anaconda can take several hours to swallow a prey. Four, six, or even eight hours are not uncommon, and an even longer time is not out of the question. Once I caught an anaconda that had recently consumed a large spectacled caiman. The snake's skin had a strong smell of rotten meat. Due to our handling, the anaconda eventually regurgitated a very decomposed caiman (I will not tell you what it smelled like!). It is unclear to me whether the anaconda scavenged and ate a caiman that was already dead or if she killed it herself, but it was so large that by the time she maneuvered and consumed it, the prey was partially decomposed. Locals later informed me that they had seen that anaconda 2 days before constricting and killing a caiman in the same location where I found her. This would indicate an extremely long handling time.

Not Usually Scavengers

Anacondas might scavenge on occasion. I have not seen it myself, but I have heard anecdotal information to this effect. At any rate, I want to take the opportunity to discuss a fascinating fact about anacondas' feeding ecology. Let's consider that an anaconda kills a deer. Since anacondas don't chew, the width of the deer, 30 to 45 centimeters (12 to 18 inches), goes intact inside the snake. Snakes that don't feed very often use a clever trick to save energy: They let their digestive system undergo atrophy so they don't have to expend energy supporting idle digestive tissue. This has been labeled "shutdown" digestive physiology.[31] So, after a big meal, the snake regenerates its digestive system so that it produces the tissue needed to make the digestive juices required to digest the recently consumed prey. In other words, during the 48 hours after consuming a prey, there is hardly any digestion. Once the snake begins producing the digestive juices, they start acting on the skin of the deer and work their way toward the internal tissues. By now the deer has been rotting in the snake's gut for at least a couple of days. So long as the deer is alive its immune system keeps bacterial proliferation at bay, but as soon as it dies bacteria in the deer's body and guts start growing at will and decomposing the deer from the inside out.

It is true that the low pH of the anaconda's digestive juices will stop bacterial proliferation instantly, but it will take no less than, say, 48 or 72 hours before the digestive juices work their way through the layers of skin and muscles. With the 2-day delay that it takes the anaconda to produce the digestive juices and the time it takes for the juices to penetrate several inches of skin and muscle, the deer flesh will have plenty of time to start decomposing before any of the digestive juices can stop bacterial growth. In other words, by virtue of consuming such large prey and having a shutdown digestive physiology, anacondas need to digest partially decomposed food. Once they had the adaptations needed to digest rotten meat, I don't see why they couldn't benefit from the eventual rotten snack if they find something really good laying around. In fact, yellow anacondas have been reported to scavenge on fishes that die in great quantities in the dry season.[32]

Studying Diet

There are some aspects of the study of anacondas that are not necessarily glamorous, and I'm not talking about shuffling about in the mud, which is actually rather cool. I have already mentioned dealing with a regurgitated meal that might have been decomposing for several days in an anaconda's stomach. The study of diet also involves a lot of playing with poop. Surprising as it might be for someone who is not a field biologist, picking up poop from your study animal is quite a valuable opportunity to study its diet, especially if you have data on the individual that passes it. There is a lot to be learned about an animal's diet by inspecting their feces. While muscle and some soft tissues are normally gone, hair, feathers, scutes, claws, and scales are normally indigestible, and we can recognize them in the feces. So, collecting poop samples is a very common procedure among biologists who want to understand the diet of an organism.

Due to their low frequency of feeding, diet samples are valuable when studying anacondas. Sometimes one finds a road-killed animal and can obtain data on the diet from dissecting it, or one catches an animal with a recent meal, but due to their low feeding frequency and because it is so risky for them to catch a prey, I don't make them vomit (Figure 5.5). Instead, I proceed carefully, depending on the size. If it is a large animal, I don't catch it. Anacondas with a big meal spend weeks in the same spot digesting it, so I can keep an eye on the snake and catch it a month or two after the meal has passed the stomach. Then I place the snake in a big bucket or holding cage to wait for the feces to pass, a week or two later.

Once a friend of mine saw a big anaconda, Courtney (377 centimeters [12.4 feet] snout–vent length, 430 centimeters [14 feet] total length, 62 kilograms [136.7 lbs]), killing a white-tailed deer. Not wanting to disturb her, and knowing what the prey was, I didn't catch her, but I kept an eye on where she was every day. I could see the bulk of her inflated stomach breaking the surface by the side of the *módulo*. For more than 2 weeks I observed the place where she was, and when I no longer saw the bulk on her stomach breaking the surface, I assumed she was about to pass it, and I decided to catch her to obtain data on her size and mark her. To my surprise, during the capture I noticed that she still had a bulk in her stomach. It was too late then to abort the capture, so I proceeded, hoping she wouldn't lose her meal. However, the handling led her to regurgitate. The entire hindquarters of the deer were still in her stomach. All the skin was gone, but about half of the deer was still undigested in the stomach, after all that time. Since then I have been very careful when I catch an animal that I know had a meal. Any reasonable estimate would have predicted that the animal was no longer in stomach after a relatively short time, but I learned differently.

Other times, the animal has eaten a while ago and with the stress of the capture it defecates. This is perhaps the most reliable source of diet information I have used. Over the years I have collected and processed a little over

200 fecal samples, and they have helped me paint a good picture of what anacondas eat.

I will never forget the expression on the face of my friend a journalist from *Reader's Digest*, who came to write an article on my research. He was quite a sport, going with me into the swamp, shuffling through the mud looking for the snakes, and even helping me handle some animals. On one occasion, though, the water had lowered so much that we were literally tripping on the snakes. There were many males and quite a few small females, so I had run out of bags to collect them. As we continued to gather them, we let them coil around our arms or legs while we focused on holding their heads. At one point I had three anacondas in my hands, two males in one hand and a medium-size female in the other, with their bodies wriggling around my arms and trunk. Jen Moore, a recent college graduate who had come to help me, was no less busy with her own load of snakes (Figure 5.6). My friend was taking pictures and carrying another small male in a bag. At that moment, I noticed that one of the snakes I had was defecating, a normal reaction to being handled.

Being in the swamp I was worried that the sample would fall in the water and would be lost. I ordered my friend in a tone of great urgency to grab the turd that the snake was letting out. Reacting instinctively to my urgent demand, he grabbed it. I then ordered him, using the same tone of voice, to place it in my pants pocket and asked him to help me remember which of the anacondas had expelled it when I could get my hands on my field notebook. This all came to me quite naturally, as in the urgency of collecting the data, I had grown calloused to what it really is. But my friend felt differently as the situation settled in his mind. Did I tell you that he was a journalist for *Reader's Digest*? At some point it might have dawned on him that he has just "delivered" an anaconda turd! He was quiet and introverted throughout dinner that night.

What's for Dinner?

One of the few times that the anacondas blow their cover from under the vegetation is when they are in the process of taking a meal. This may be the reason that opportunistic documentation of feeding events is not uncommon in the literature, even if finding anacondas is not easy. Some reports include tamarind monkeys, spectacled caimans,[28] peccary,[30] and a few more, including three events of predation on other snakes,[33–35] but before my study no systematic and comprehensive study of the anaconda diet had been done with data from a single location describing the gamut of prey anacondas feed on.

In the anaconda's menu we can find a wide variety of terrestrial vertebrates. I have identified over 27 different species. Most of the stomach contents contained only one specimen. Approximately half of the diet is birds, with large mammals and large reptiles making up the other half in nearly equal proportions. I found only one record of a fish, and the consumption of this item is questionable due to the fact that the snake did not actually swallow the animal (see later in this section). The lack of fishes in the diet of anacondas is surprising given the high density of fishes and their aquatic habitats. This finding is also surprising even if there are isolated findings of anacondas feeding on fishes in other locations in South America (Figure 5.7). Anacondas seem to be opportunistic predators that will take any prey they can kill and swallow (Figure 5.8).

As my data point out, newborn animals feed mostly on birds from the very early ages. One of the females that I had in an enclosure to study reproduction gave birth in an outdoor cage. I collected all the babies in the enclosure but failed to find one. About 3 weeks later, I found that the neonate I had missed, still in the enclosure, that had eaten a free-ranging small passerine bird (*Phacellodomus rufifrons*). Other neonates were found preying on jacanas (weighing 70 grams [2.5 ounces]) and other small wading birds (Figure 5.9). Judging by the habitat newborn snakes use and the timing of their birth, it is likely that they also feed on newborn caimans (weighing approximately 40 to 50 grams [1.4 to 1.8 ounces]) and side-neck turtles (*Podocnemis vogli*, 10 to 15 grams [0.4 to 0.5 ounces]), but no record of these has been found so far. However, the sample of neonates is not as large as we wish. Theoretically, they also could catch amphibians. This would not be easy to identify in the feces for their lack of hard tissues, but the neonates are born at the end of the wet season when the numbers of amphibians decline. Amphibian numbers do not recover until 6 months later, so newborn anacondas probably do not rely on them for survival. I found no evidence of anacondas feeding on any amphibian at larger sizes. Although their remnants would not appear in the feces of larger snakes either, I never witnessed a predation attempt, found them in the diet of roadkill snakes, or found them regurgitating amphibians during captures. To the best of my knowledge their predation on amphibians is a rare event, if it happens at all.

Despite the relatively clear trend of the diet showing preference for birds, like everything else regarding anaconda biology, the difference between the sexes is striking and cannot be ignored. Males eat mostly birds while females consume mammals, birds, and reptiles in about equal proportion (Figure 5.10). However, the difference in diet seems to be associated with the size difference. Smaller female anacondas feed heavily on birds, just like males do, and gradually include reptiles and mammals in their diet as they grow larger (Figure 5.10). Large anacondas feed almost exclusively on larger mammals and reptiles (Figure 5.11).

Only 30% of the animals for which I have diet data are males. This is surprising, mostly given the fact that males outnumber females more than 2:1. This strongly suggests that females are a lot more likely to be found eating, or after a recent meal, than males are. This could be due to two non-mutually exclusive explanations. The act of capturing a meal makes the anacondas a lot easier to spot, which biases the representation of female diet in the sampling. This may also be part of the reason that females are overrepresented in the general sampling. Also, females may be feeding a lot more often than males, which means they are more likely to be found during or after feeding. This higher feeding frequency will also explain their higher growth rate (see Chapter 4).

Meal Size

In some cases, anacondas go after large prey, tackling animals that may be larger than themselves. After tagging Olivia I had a chance to test experimentally the popular legend that anacondas swallow a deer butt-first and let the neck rot before finishing the swallowing process. One day I found Olivia next to a recently killed full-grown white-tailed deer with a full rack of antlers. Olivia was apparently unable to swallow it due to the deer's antlers. I could recognize the tooth marks on the deer's foreleg where she probably grabbed the deer to throw her powerful loops around him. On close inspection I could see the tooth prints, all four rows of the snake's upper jaw, marked on the deer's face all the way to the point where the antlers began. Apparently, she attempted to eat it, head-first but was unable to swallow it past the antlers. Olivia lay in the water close to it for several days, and eventually vultures ate the deer. Olivia never attempted the butt-first approach and instead let 70 kilograms (154 lbs) of meat go to waste after expending all the effort of capturing and killing the big prey.

It had taken almost three decades to put that question to rest in my mind, but I finally had an experimental answer: No, anacondas do not eat the dear butt-first and wait for the neck to rot. The adult deer that anacondas feed on could be antlerless females or males that have recently shed their antlers. When a hungry female anaconda finds a deer at the edge of the water, she probably does not give a lot of extra thought to whether it has antlers. It's probably not convenient for her to risk letting the animal escape on the off-chance that it might be a male deer with a full rack.

The first measurement I had from Olivia recorded her weighing 43 kilograms (94.6 lbs). A full-size male deer, with an average mass of 60 kilograms (132.3 lbs), represents a whopping 140% of the snake's mass. A prior meal that Olivia took was a 2-meter-long (6 feet) spectacled caiman with an estimated mass of 38.5 kilograms (85 lbs), which represents nearly 90% of her body mass. However, these proportions perhaps overestimate the relative size of the prey because Olivia was very skinny at the time of the capture. An anaconda of Olivia's size would weigh in excess of 70 kilograms (154 lbs). However, the deer or the caiman still represents a substantial meal for the snake's size. Imagine a person eating a comparable amount—say, 154 lbs of meat (2,121 hamburger patties)—at one sitting!

Due to their extraordinary sexual size dimorphism, the size of prey an anaconda may attack varies considerably between the sexes. Mammals and reptiles are the largest prey anacondas feed on (see Figure 5.12). Males attack smaller prey than females in absolute but not so much in relative terms in the same range of size. Relative prey size in males is somewhat smaller (20.9%) than females (31.1%), but this does not seem to be a strong difference within the same size brackets (Figure 5.12). But because their size is so different, females consume substantially larger, and more dangerous, prey than males in absolute terms. Due to their reproductive biology—females carry the neonates for 7 months, during which they do not eat—females incur a very high cost of reproduction. It seems that this high cost of reproduction for females results in females needing to acquire more food, leading them to attack larger and more profitable prey. In fact, the difference in diet composition between males and females seems strongly linked with the female's reproduction. The diet of males and females is nearly identical until females reach reproductive size (around 2.1 meters [6.9 feet] snout–vent length).

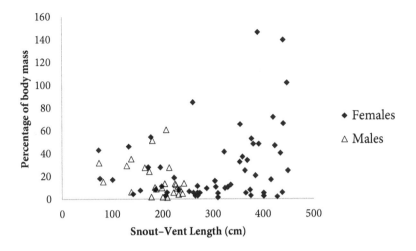

Figure 5.12 Relative prey size in green anacondas in the Venezuelan llanos. Females take larger prey as they reach adulthood.

Then they consume a lot less of the very lean birds and focus on eating mammals and reptiles, which contain a lot more fat.

It is also surprising the proportion of ducks found among females. The only birds we find regularly among reproductive females are ducks, but not any kind of ducks. In the anaconda's diet, Canadian blue teals (*Anas discor*) outnumber whistling ducks (*Dendrocygna* spp.) approximately 9 to 1. This is surprising because the proportion of ducks in the llanos during the dry season is almost the opposite of that: We can see large flocks of whistling ducks and very few Canadian blue teals. This presented another enigma to me since if you pick ducks at random, you are a lot more likely to pick a whistling duck than a Canadian blue teal. Why do females feed on the rare ducks more than the more common ones?

My first impression, a Venezuela-centric one, was that the Canadian duck was not as fit to avoid the anaconda's predation as the local one. However, Canadian blue teals are only Canadian by name; they have evolved for tens of thousands of years visiting anaconda territories, and evolution might have acted on them as well to help them avoid anacondas as predators—although it is true that the "Venezuelan" ducks experience anaconda predation throughout the year and the "Canadian" ones suffer it for a shorter time. It is also true that a strong selection pressure such as predation would select for ducks that can avoid anacondas, Canadian and all as they may be. As far as I can tell, the main difference between individual whistling ducks and Canadian blue teals is that the latter migrate thousands of miles and gather a thick layer of blubber around their body to supply the energy needed for the long flight (Thanks to Patrick Weatherhead for the insight). Many of my data come from the months of March and April, when the snakes are easy to find. This is also the precise time when blue teals are preparing for their long migration and gathering the energy for the long flight. With this in mind, it turns out that it's rather smart for anacondas to target the fatty and more nutritious Canadian blue teals for their meals. As far as anacondas are concerned, each blue teal may well be a duck wrapped in a thick layer of bacon ready to be swallowed!

Cost of Predation: How Much Is Supper?

It is not surprising that female anacondas take larger, more nutritious prey due to the metabolic needs that reproduction imposes on a female. Female anacondas do not feed during pregnancy, so they need to gather enough fat before they go into a reproduction event. However, this switch in diet comes with a price. Feeding on large caimans is obviously dangerous due to the tough weaponry that caimans have and their potential to hurt the anacondas. Capybaras, cute as they are, are rodents after all (Figure 5.13). Imagine a rat the size of a Rottweiler. Now scale the rat incisors to the size of the dog, and that's the weaponry that anacondas have to deal with every time they attack a capybara. Consider now that capybaras live in social, family groups that often protect each other in situations of danger. Now we have a pack of 10-plus Rottweiler-sized rats that the snake has to deal with! It is not uncommon to find female anacondas with large wounds, likely inflicted by capybaras (Figure 5.14). Even less well-armed prey can represent a challenge by virtue of their size. When consuming a very large prey, an anaconda might risk asphyxia as the bulk of the prey can prevent the animal from breathing.

I documented several occasions where anacondas were wounded or even killed by their prey.[4,28] One of the radiotagged snakes, Susana (455 centimeters [15 feet] total length, 46 kilograms [101.4 lbs]), was found floating in the middle of the pond at 6 a.m. By 8 a.m. she had not moved, and I touched her, confirming that she was dead. Her head was missing, but no other injury was apparent upon the physical examination and necropsy.

The next day a young capybara weighing approximately 2.5 kilograms (5.5 lbs) was found floating on the pond. It had been dead for approximately 24 hours and showed clear scratches and an anaconda's tooth marks in the middle of its body. It also had a dislocated spine at the cervical level. I surmised that Susana attacked it and then she was attacked by the relatives of the baby capybara that bit her in the neck, killing her. Piranhas and other scavenger fishes might have eaten Susana's head. Attacks of capybaras on anacondas are relatively common, as evidenced by scars and wounds on several female anacondas that matched the size and shape of capybara teeth (see Figure 5.14).

I once came close to witnessing one of these defensive attacks, but my presence probably altered the outcome of the event. I was wading in a river looking for snakes and heard the squealing of a baby capybara ahead of me. As I ran to the source of the noise, I saw a herd of capybaras on the riverbank running in disorder in all directions; however, a female remained standing, looking in the direction the squealing was coming from, some bushes at the water's edge. The female looked at me as I approached and looked down at the bushes again. She repeated the movement several more times as the distance between us shortened, as if assessing what she should do. When I was apparently too close for her to stay there any longer, she jumped into the water and reunited with the rest of the herd. In the bushes where she was peering, I found Zuca (female, 504.7 centimeters [16.6 feet], 62.5 kilograms [137.8 lbs]) constricting a juvenile capybara (estimated mass 15 kilograms [33 lbs]). I did not capture her so as not to interrupt her feeding, but I could identify Zuca by the location, size, and distinctive marking and scars that she had next to her neck. Since I was approaching, I probably deterred the female capybara from attacking the snake, but of course, there is no way to know.

Another predatory event that brought complications to the predator involved a rather small prey but a tough one. I was informed by some local people that a large anaconda was seen eating a large llanos side-neck turtle. Two weeks later I saw Francies (485 centimeters [16 feet] total length, 61.3 kilograms [135 lbs]) in that same spot with very loose skin around her mouth and neck. After capture, I saw that the skin and flesh were torn more than 20 centimeters (7.9 inches) from her mouth to her neck. I collected measurements of the animal and put her back in the field. I kept an eye on her regularly and saw her basking often in the same place. Basking for long periods is often associated with some infection. Snakes give themselves a fever, which helps them fight infections. Two days later Francies died.

On the necropsy I found epidermal scutes of *P. vogli* that matched the scutes of a turtle shell 20.3 centimeters (8 inches) long. The only explanation I can think of is that the turtle was too large for the snake, and it perhaps was in the wrong position. The edge of the turtle shell might have torn the esophagus and skin of the snake as she was swallowing it (Figure 5.15). It is surprising, though, that she didn't stop swallowing when her skin first ripped.

This observation makes you wonder what the mental state is of an anaconda while it's swallowing a large prey. When looking at an anaconda that is engulfing a large prey, one gets the impression that the distension of the skin and movement of the jaws must be painful (Figure 5.16). It would certainly be painful for a human being, but it must not be the case for an animal that has been shaped by natural selection to do just that. However, it seems like the pain receptors of Francies were not fully engaged while she was swallowing the turtle. It was as if some natural analgesic allowed her to perform such a feat of stretching but also prevented her from feeling when her flesh was tearing. This may be similar to the analgesic effect that, say, adrenaline has in the human body. This is the reason that when we suffer an injury while exercising, we often don't feel the full intensity of the pain. Is there a similar process happening in anacondas that take a large meal? Unfortunately, cognition in snakes is a very incipient science, and it will be a long time before these ideas can be tested experimentally. It remains puzzling why a snake would continue swallowing a prey that was damaging her body to such an extent.

The only instance of an anaconda preying on a fish that I have witnessed was a male anaconda (268 centimeters [8.8 feet] total length, 10.7 kilograms [23.6 lbs]) swallowing a catfish (*Pseudopimelodus apurensis*; 29 centimeters [11.4 inches] total length, 425 grams [15 ounces]). This fish is characterized by a wide head bordered by long sharp spines. These spines had punctured a large hole through the snake's esophageal wall, muscle, and outer skin (Figures 5.17). The fish, still alive, was biting the snake's esophagus and was held in place by some of its spikes that were stuck in the snake's skin. I caught the snake and extracted the catfish with very little effort through the hole it had made through the snake's esophagus and skin.[4] The snake was kept in captivity with no treatment until it healed completely. About 2 months later the snake ate an adult iguana that was placed in the enclosure, showing that it had fully recovered. It would be no surprise to learn that this anaconda lost its appetite for seafood after this encounter! The fact that anacondas do not feed more on fishes comes as a surprise considering the abundance of the fish as potential prey. Perhaps free-swimming fish are too fast for the anaconda to catch in open water, and those benthic fishes (like this catfish) often have spines and weaponry that prevent them from being easy prey. This is the only case I witnessed of a male suffering a serious injury while trying to consume a prey. The trend is for females to be the ones that get wounded. It shows that feeding may always be a challenge for anacondas.

I collected thorough information on scars and wounds in a subsample of my data. Many animals displayed a large number of scars on the skin, apparently a result of attacks from potential predators or the defensive efforts of prey. Females had scars more frequently (35 of 38 animals) than males (25 of 56 animals), and the number of scars found on females was higher than those found on males (Figure 5.18). This seems to be correlated to the larger size that females reach. Animals that grow older have time to accumulate more scars; thus, larger animals have more scars both because they have endured more predaceous attempts and because they attack larger prey. The lower number of scars that males typically have may be due to the fact that they attack smaller and less dangerous prey. Notice that Figure 5.19 almost mirrors Figure 5.12 that shows the relative prey size. Larger prey seem to represent a clear and present danger for the females.

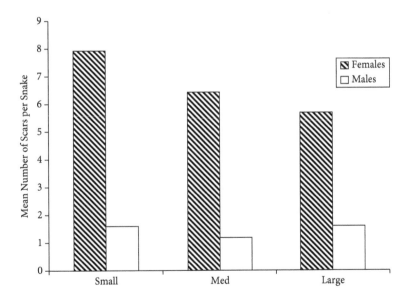

Figure 5.18 Scars found on wild anacondas in the llanos. These scars are often the result of predation events against well-armored prey. Females of all sizes show larger numbers of scars.[4]

Reproduced with permission Eagle Mountain Publishing, LLC.

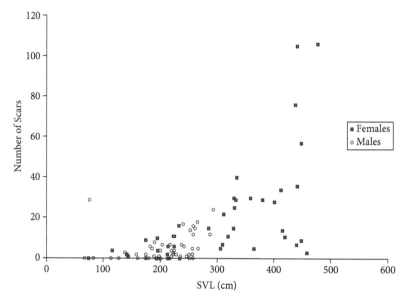

Figure 5.19 Scatter plot of number of scars found in males and females in the Venezuelan llanos. Adult females show a substantially larger number of scars than young females and males.[4]
Reproduced with permission Eagle Mountain Publishing, LLC.

An alternative explanation for the higher number of scars in females than in males is that males face a higher mortality than females while tracking females during courtship and mating and do not live long enough to gather a large number of scars in the skin. However, this second scenario would involve high mortality in males and a female-biased sex ratio, which is not consistent with the demographic data. In fact, the survival data show just the opposite, with females having higher mortality than males. Actually, the high risk of dying in females associated with attacking larger prey would be another contributing factor to the male-biased sex ratio documented in Chapter 4.

Predators should avoid prey that can injure them since recovering from wounds can be so costly that the benefit of the meal is largely spent on it. Why anacondas attack prey that can, and do, inflict serious injuries on them is a question that is very relevant to the ecology of the species. Aren't there other prey available? The abundance of waterfowl and other relatively "safe" birds in the llanos does not seem to support this hypothesis (Figure 5.20). However, birds are a fairly lean prey item since flying imposes weight constraints on them, and they might not pay for the metabolic cost of digestion. Furthermore, catching and digesting a bird might prevent the anaconda from catching another, more profitable prey. Thus, smaller prey may be dropped from their diet as they are simply not worth it. Reticulated pythons (*Python reticulatus*) show a similar switch in diet in approximately the same size classes that anacondas do, with an associated decrease in the feeding and breeding frequency of larger individuals.[36]

Recall that anacondas likely have a shutdown physiology where every time a snake eats a prey, it likely has to invest a lot of energy in regenerating its digestive system. Consuming a prey that will not provide enough energy to start the system and still produce a payoff might result in a negative energetic gain, a loss. It is likely that larger anacondas prey mostly on mammals and reptiles (and very fatty ducks) because these prey have more fat reserves and are energetically more profitable, even if some might involve a high risk of injury.

A Man-Eater?

Of course, people who wonder about anacondas as predators end up, sooner or later, asking whether anacondas are "man-eaters." Let me start out by stating that because humans arrived in South America only in the last 15,000 or 20,000 years, there is no evolutionary history of anacondas, or any South American predator, eating humans. This is different from African pythons, which have a long history with encounters with humans.[37] Anacondas did not evolve preying on humans or anything that looks like a human, since there are no apes on the continent. Thus, anacondas are clearly not natural predators of humans. They do prey on other primates, but there is a substantial size difference between humans and South American primates.

Predators often evolve feeding on certain prey for which they may have adaptations to recognize and are well equipped to attack and subdue. Unless a predator has this innate drive to feed on certain prey, it will normally avoid any new animal it encounters for the first time. For starters, the new animal could be a predator itself, or it may have dangerous weaponry that can injure the predator, so it is better to avoid the unknown. Optimal foraging theory predicts that if a potential prey becomes very abundant and the encounter rate with the predator is high enough, the predator will start considering it as a prey.[41] If there are enough of them, it will be worthwhile for the predator to invest in learning how to deal with a new prey and to take the chance of being injured in the process. So, after a number of encounters, the predator should start considering the unknown animal as a potential prey and perhaps try its luck with it. Because anacondas live in the swamp, where humans do not often go, the encounter rate between humans and anacondas is low. This may be the reason that anacondas are not commonly reported attacking people and that there is no substantiated predation by anacondas on humans.

While nobody doubts that anacondas are strong enough to kill an adult person, some people wonder if they would be able to pass the shoulders of the person. I have not done particular studies about the anaconda's jaw mobility, so I don't have novel information to offer on this area. However, anacondas can consume a capybara that is as big as 46 centimeters (1.5 feet) across. Recall that the shoulders are mobile and contain joints that could be broken or disjointed. I strongly believe that anacondas would not give up 55 to 80 kilograms (150 to 180 lbs) of meat for a rather small obstacle such as the person's shoulders after they have done all the work of hunting, stalking, constricting, and killing. At least this is what anacondas seem to believe. I have witnessed two occasions in which anacondas tried to eat one of my assistants. Their lack of success had nothing to do with gape size but rather with the awareness the "prey" had of the snake in the moment of the attempted predation, and external help received on one occasion.[24]

The first attempt was by a large female (Lina; 5.04 meters [16.5 feet] total length, 54 kilograms [119 lbs]) that had had a serious mouth infection at the time I captured her and implanted a radiotransmitter in her. Two months after implantation, I found Lina at the edge of the water in an area of the river without vegetation. Because of her mouth infection I stayed there longer than I needed to collect the telemetry data, trying to assess the status of her infection, but the murky water prevented me from making a reliable observation. Lina stood at the water's edge tongue flicking for a long time, clearly seeing and smelling me. Nothing happened at the time, but events that happened later make me think that she might have been sizing me up.

The following week my coworker María (156 centimeters tall, 56 kilograms [5- foot- 1, 123.5 lbs]) and I followed Lina's transmitter signal to collect data on her mobility and also tried to observe her to assess her infection. The snake was in a shallow channel, approximately 80 centimeters (31.5 inches) deep, which was partly covered by emergent aquatic vegetation (*Eleocharis* sp.). María walked the 50 meters (54.7 yards) between the boat, where I stayed, and the spot where the signal came from to pinpoint the location so she could collect data on where Lina was. At my prompting, she stayed a bit longer trying to assess her mouth infection without being able to see it. Suddenly the snake emerged from the water and struck, grabbing María by the knee. Fortunately, her pants tore and the snake did not get a firm hold to drag María into the water. The snake immediately struck again with her mouth open to about 180 degrees, this time at the level of María's waist. However, her prompt retreat resulted in an unsuccessful attack. The good news is that Lina's mouth infection had healed quite well!

A week after this incident I found Lina again on the same river. She was next to the bank on the other side of the river some 15 meters (16.4 yards) away from my side. I started collecting data and taking notes on her location, and to my surprise when I reached into the river to measure the water temperature I saw Lina's head right by my ankle, in what looked like stalking. Apparently, she saw me from the other side of the river and dove under the water to come out right at my ankle. This shows both a great sense of underwater navigation and a lot better eyesight than people often concede to snakes. After seeing the attack on María I wondered if the prior week she hadn't been sizing me up while I was trying to assess her infection. Of course, at 1.78 meters and 83 kilograms (5- foot- 10, 187 lbs), I was a much more challenging prey than María!

Determined to document experimentally if Lina would attack a prey 160% her body mass (me), I stood still where I was. I made sure my hand was ready to grab her by the neck should she seize my ankle (the closest part for her to grab) and waited for her to make the next move. The standoff might have lasted some 10 or 15 minutes with me waiting immobile, with my heart pounding in my chest wondering if this was such a good idea, and her tongue flicking, intensely, seemingly sizing me up as a prey. She eventually went underwater and left. She swam down the river where, some 200 meters (218.7 yards) away, a mare and a colt were frolicking in the river. Perhaps she was deterred by my size, or perhaps it was my eyes intensely looking at her (predators seldom attack a forewarned prey). In retrospect, I'm not sure it was such a good idea to invite an attack by a 5-meter-long (16-foot-long) snake. After seeing how they constrict their prey and knowing that the constriction is intended not just to asphyxiate but also to pull and twist, and break, the spine of their prey, I am very glad Lina decided to stand down!

The other event involved Ed George (1.74 meters [5- foot- 9], 57 kilograms [125.7 lbs]), a photographer friend of mine who, not for nothing, earned the nickname "Gringo Loco" in my study site. We were doing a documentary for *Discovery*, and the film crew was eager for us to find a big anaconda. Since they had two cameramen, Ed joined the search team to increase the odds of finding a snake while the other cameraman filmed the action. We were looking for snakes in a river covered by aquatic hyacinth (*Eichhornia* spp.). We walked past a snake without detecting her, but she did notice us. The cameraman who was filming us saw the head of Penelope (as we named her afterward) appearing right behind Ed as he walked by. In the footage it is possible to see Penelope following Ed, tongue flicking intensely, for approximately 2.5 meters (8.2 feet), while raising herself up to 25 centimeters (8 inches) above the aquatic vegetation. She was following so closely that Ed kicked her on the chin with his heel when he picked up his feet while walking. It took a few seconds for the cameraman to warn Ed about a large snake stalking him (perhaps hoping to get the shot!). Eventually he did warn us, and we turned around to grab the snake. I managed to grab Penelope by midbody just as she struck at Ed, who in turn jumped out of reach. Both actions, my pulling the snake backward and Ed jumping out of reach, made the snake fail and snap into the air. On catching and subduing Penelope (whom I named after Ed's daughter), she measured 445 centimeters (14.6 feet) total length and weighed 39 kilograms (86 lbs). She looked healthy but very thin. I surmised that she had given birth the prior year (perhaps in November) and had not eaten since. Since it was March, and females do not feed during pregnancy, it is possible that Penelope had not eaten anything for a whole year. Not surprisingly, she was ready to attack a prey 146% of her body mass!

There is no doubt in my mind that both of these attacks were predatory in nature. In the first instance, the following evidence suggests that the snake must have been foraging when she attacked María:

- Lina had not eaten during the 2 months she had been radiotracked, and probably longer, due to an oral infection.
- Eight days before the incident, the snake thoroughly tongue-flicked at me and did the same the following week.
- The following week I found Lina with an engorged stomach, evidence of a big meal, so Lina was definitely foraging during the time she attacked María.
- Lina's attack was unlikely to have been defensive. By the time she struck at María, she had not been caught or bothered. The water was deep enough that she had plenty of opportunity to slip away unnoticed. In my experience of catching hundreds of anacondas of all sizes, I have found that large individuals are very unlikely to show any defensive behavior when

Figure 1.1 The anaconda, the largest snake in the world, contends with some pythons for the title of the longest one, but no one comes even close to it in mass and bulk. Refugio de Fauna Rio Guaritico, Hato El Frío. (Photo Philippe Bourseiller)

Figure 2.1 John Thorbjarnarson (middle) and María Muñoz (left) were critical figures in my research (Jesús Rivas on the right), certainly in the first year of the study. (Photo Bill Holmstrom)

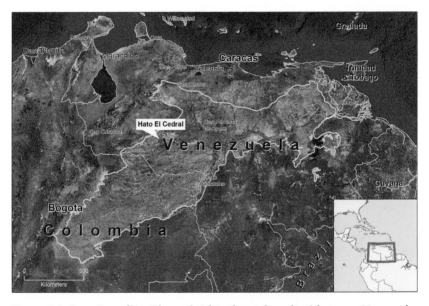

Figure 2.2 Location of Los Llanos (within the pink outline) between Venezuela and Colombia. The study site proper was right under the first "e" of Venezuela. (Map by Joe Zebrowski)

Figure 2.3 As the dry season gets stronger, bodies of water shrink, which leads to concentration of wildlife around them. Rio Matiyure, Hato El Cedral. (Photo Jesús Rivas)

Figure 2.4 El Cedral was chosen for its good internal roads and easy access, plus its high anaconda density. (Photo Robert Caputo)

(a)

(b)

Figure 2.5 Seasonality in the llanos is extreme. The pictures show the same landscape at different times of the year. (Photo Robert Caputo)

Figure 2.6 Road view of El Cedral. The uplifted road from north to south represents a dam in a landscape that tilts west to east. These accidental dikes hold water on their west aspect. These created an important habitat for the study: *módulos*. (Photo Robert Caputo)

Figure 2.7 Llaneros work in the llanos on horseback and were very knowledgeable about anacondas. Their help was invaluable when I started the project. (Photo Tony Crocetta)

Figure 2.8 Ramón Arbujas was one of the llaneros who helped the most with his knowledge about anacondas and how to find them. Here Ramón explains to us how to find snakes under the dry mud with a stick. Left to right: María Muñoz, Jesús Rivas, Ramón Arbujas, John Thorbjarnarson. (Photo Bill Holmstrom)

Figure 2.9 Anacondas blend in with the aquatic vegetation. Even a large animal may be inconspicuous due to its cryptic coloration and ambush hunting style. (Photo Tony Rattin)

Figure 2.10 Searching team wading through the swamp feeling for snakes under the vegetation with our feet and poles. Left to right: Neil Ford, Lars Holmstrom, Qing Qiu, Jesús Rivas, and Gabriela Jimenez. (Photo Bill Holmstrom)

Figure 2.11 The spectacled caiman is a frequent neighbor (both as predator and prey) of the anacondas in the swamps. Their habitat preference being very similar to those of anacondas led to many accidental encounters with them in the field. (Photo Jesús Rivas)

Figure 2.12 Orinoco crocodile, one of the most endangered species of crocodiles in the world, once was abundant in the Venezuelan llanos. Skin trade as well as fear from the local inhabitants brought it to the brink of extinction. El Cedral is one of the sites where crocodiles have been reintroduced lately as a part of conservation efforts, and they seem to be making a comeback.

Figure 2.13 Finding anacondas under the mud was always a big adrenaline rush! Gabbing the head was just the beginning. (Photo Paul Calle)

Figure 2.14 Handling snakes often involved physical struggle with the animals until we rendered them exhausted. (Photo Robert Caputo and María Muñoz)

Figure 2.15 Anacondas tire quickly due to their anaerobic metabolism. After a few minutes of struggle, we were tired and panting but were still able to move; the snake was not. (Photo Tony Crocetta)

Figure 2.16 Kicking off my shoes while searching for anacondas proved to be a handy way to avoid foot rot and to more easily recognize animals under the water. (Photo Tony Crocetta)

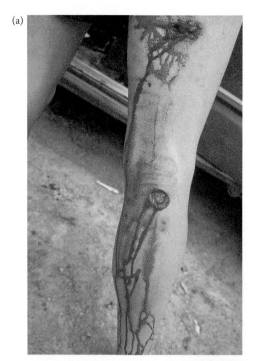

Figure 2.17 Close-up look at piranha (*Pygocentrus cariba*) dentition. While a bite can be a serious injury, the true danger of piranhas has been grossly exaggerated. However, accidental bites may take place on occasion. (Photo Paul Rafaelle)

Figure 2.18 Anaconda swallowing a white-tailed deer. Notice the extension of the jaws and design that allows the snake to swallow large prey. (Photo Jesús Rivas)

Figure 2.19 Anaconda making a loop with its body to protect the head. In this picture the snake has pulled out its head to the side, still presenting the muscular bulk as a target for aggressors. Both hands of the handler have been caught in the "evil loop," which would render the handler helpless if he does not free his hands. (Photo Sarah Corey-Rivas)

(a)

(b)

(c)

Figure 2.20 The "evil loop" can be controlled if the handler focuses on the first meter of the snake. On occasion a snake may choose to raise herself, trying to free her head during the struggle. This is the tipping point since that effort would in all likelihood render her exhausted. (Photo Carol Foster)

Figure 2.21 Given the snake's low endurance, it is possible to control even a large snake without requiring excessive strength. (Photo Paul Rafaelle and Tony Crocetta)

Figure 2.22 Bob Caputo, a photographer from *National Geographic*, holding an exhausted snake, Joan. Joan had been captured every year for 6 years and was not as worried about the person holding her as she was about me, taking the pictures. (Photo Robert Caputo)

(a)

(b)

Figure 2.23 Once muzzled, the snake behaved calmly, allowing easy data collection. Measuring snakes by following the dorsal medial line offers the best way to determine the length of the snake. (B) Group of students, helpers, and collaborators. Left to right: Katz Treseder, Jack Hoopia, Rose Peralta, Simon Treseder, Sarah Corey-Rivas. (Photo Justin Saiz)

Figure 3.1 At first we needed a lot of help from volunteers to subdue and restrain the big anacondas. Left to right: Reneé Owens, Mark Masio, Richie Zirelli, and Jesús Rivas. (Photo Robert Caputo)

(a)

(b)

Figure 3.2 Looking for anacondas in a dry river I spotted the corpse of a large female only to discover a carpet of them breeding in the mud. In the background of the first picture (a) is one of the German helpers who joined our team. In the second picture (b) María and I are trying to hold on to a carpet of anacondas in the river. (Photo Bill Holmstrom)

Figure 3.3 Securing anacondas on the riverbank. This was the first mating aggregation I ran into. Two German helpers, María, and I (hat in hand) holding the ropes. (Photo Bill Holmstrom)

Figure 3.4 Looking for anacondas in the swamp during the wet season proved to be quite a challenge. In this image is Cesar Molina, one of my good friends who came to assist me in the fieldwork regularly. It looks like he would rather NOT find an anaconda in that moment! (Photo Robert Caputo)

Figure 3.5 Hunting party looking for green anacondas in the Venezuelan llanos. Chuka was a feral dog that I adopted in the area. She was young enough that I could tame her and make her my pet. (Photo Robert Caputo)

Figure 3.6 Looking for anacondas in the wet season often proved to be disheartening because of their extremely high mobility. (Photo Tony Crocetta)

Figure 4.1 Captures of large snakes. After hours looking for snakes it was always great to finally find some animals under the mud. (a) With Ed George capturing Ashley. (Photo Tony Crocetta) (b) With Victor Delgado in Rio Matiyure. (Photo Ed George)

Figure 4.3 The first snake I caught in Hato Flores Moradas. Left to right: Kako, Jesús Rivas, and Walter Timmerman. (Photo John Thorbjarnarson)

Figure 4.4 Large female anaconda killing a capybara. Anacondas are far more conspicuous when they are in the act of killing or digesting a prey than when they are ambushing or resting under the aquatic vegetation. (Photo Carol Foster)

Figure 4.5 Predators of juvenile green anacondas. Neonatal anacondas rely on crypsis to survive. Once their cover is blown they are easy prey for a variety of predators. (Photos Ed George)

Figure 4.6 A strong dry season might become deadly for anacondas if they are caught by the intense sun out of the water. (Photo Tony Crocetta)

Figure 4.7 Anaconda hiding under the mud sitting out the dry season. (Photo Ed George)

Figure 4.8 Sometimes I found anacondas dead in the field with a relatively mysterious cause of death. Later analysis revealed various reasons, including parasites or other diseases. The life of an anaconda is always a dangerous one, even after outgrowing most predators. (Photo Ed George)

Figure 4.9 María Eugenia (356 cm [12 feet] TL, 20.5 kg [45 lbs]) had a bad eye infection. We treated the infection externally, but the problem went far deeper than we could cure using a field procedure. María Eugenia lived at least another year with the infection. She reproduced and killed at least two prey items during that time. (Photo Paul Calle)

Figure 4.10 Neonatal anacondas live among the water hyacinth, often blending in well with the background. Just like adults, they prey on smaller vertebrates that approach the water's edge. (Photo Carol Foster)

Figure 5.1 Black-crowned night heron (*Nycticorax nycticorax*) being caught by an anaconda. The strike was so fast that I never saw it being grabbed. (Photo Jesús Rivas)

Figure 5.2 Anaconda killing a turtle. Drowning the turtle by holding it underwater for a long time while the snake breathes is the simplest way to kill an aquatic turtle since the snake cannot stop its heart by squeezing and does not have any other way to prevent it from breathing. (Photo Tony Crocetta)

Figure 5.3 (a) Anaconda eating a deer in the swamp in Venezuela. (Photo Jesús Rivas) (b) Sometimes the constriction of the snake is so strong that it may break the ribs of the prey. (Photo Paul Calle)

Figure 5.4 Green anaconda constricting a spectacled caiman (*Caiman crocodilus*). Notice the angle on the back of the caiman. The sharp angle suggests that the spine might be broken. (Photo Tony Crocetta)

(a)

(b)

Figure 5.5 Very large animals can consume very large prey. While digesting a meal, snakes are very vulnerable. With the large bulk in their stomach, they cannot swim well or escape from or fight off a potential predator. (Photo Carol Foster)

Figure 5.6 Biting off more than we could chew, with Jen Moore. After a sudden drop of water level, we ended up finding more anacondas than we could carry. (Photo Paul Rafaelle)

Figure 5.7 Rare event of an anaconda eating fish. This picture was taken in Bonito, Mato Grosso do Sul, Brazil, but even there an anaconda eating fish is a rare sight. (Photo Luciano Candisani)

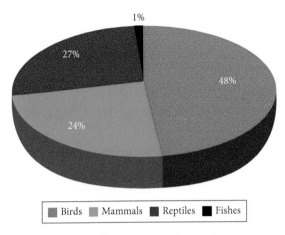

Figure 5.8 Diet of green anacondas in the Venezuelan llanos. Birds seem to be important players in their diet, especially among the young ones.

Figure 5.9 Anaconda preying on wattled jacana (*Jacana jacana*). A meal is such a coveted opportunity that the small snake is reluctant to let it go even after being disturbed and moved out of the water. (Photo María Muñoz)

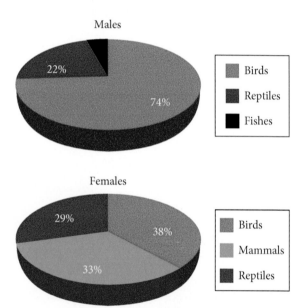

Figure 5.10 Number of prey items of different categories found in the diet of green anacondas.

disturbed. Recaptured animals are, if anything, even more skittish than naive ones (before handling) and try to escape as soon as they detect the proximity of the researchers. Thus, the proximity of the researcher is unlikely to have induced a defensive strike.

- María was well within the range that an anaconda the size of Lina can catch, as anacondas have been reported to feed on prey that weigh as much as they do.[24]
- Lina was being followed by telemetry, thus enhancing the number of times she encountered a human being. The artificially high encounter rate produced by the telemetry study might have exceeded the threshold of encounters that makes a potential prey item profitable, despite the high risk of attacking a large and potentially dangerous prey.[38]

The second event was performed by a large animal that was fairly thin for her size. Analysis of the video shows that Penelope performing predatory tongue-flicks directed toward Ed; she was following him in an obvious stalking fashion. She was probably in a large energetic deficit, so taking the risk of attacking a large prey was warranted. The prey/predator mass ratio of this event (146%), while impressive, falls within the reported prey/predator ratios for other snakes[24] and is not unheard of among anacondas. There is no doubt that both snakes were strong enough to subdue and kill the people they were stalking and, as I observed earlier, the width of a person is hardly a challenge for a hungry anaconda after she has gone to the effort of stalking, coiling, and killing the prey.

Now, does this mean that anacondas are man-eaters? Not necessarily. Both attacks were on people who were looking for anacondas in places where people often do not walk. When we look for anacondas, we wade on the swamp for 8 hours a day or more, which increases tremendously both our exposure and our encounter rate (that is why we do it!). Locals, on the other hand, avoid these habitats as much as they can, and when they must cross them, they normally go on horseback and swiftly. Although anacondas are not man-eaters by nature, they are generalists and will take any prey they can subdue and swallow, including humans. As humans encroach more on anacondas' habitats, encounters with snakes are expected to become more abundant, and thus the odds of human predation become higher.

6

The Largess of Motherhood

Fecundity and Female Reproductive Strategy

Federmann ... wandered out into the Llanos to the banks of a mighty river. Since there were various signs that the region at other times had been densely populated, Federmann wished to ascertain the cause of its present desolation. He learned from the several captured Indians that in the river there lived an animal so carnivorous and voracious that it had eaten many of the inhabitants. The rest had abandoned the site and fled to a remoter section to escape the ferocity of so deadly an enemy. Federmann and his soldiers considered this statement true because by night they had heard the formidable bellows of the wild beast. Some even said they had seen it and affirmed that it was a species of serpent of terrifying corpulence.

From *The Conquest and Settlement of Venezuela*[39, p. 56]

When I asked local people about anacondas, I frequently heard about large snakes, often of fantastic sizes. At first I rushed to the place where a leviathan-like snake had been sighted, only to learn that the snake that was reported to be 15 meters (49 feet) long was only 3.5 meters (11.5 feet) or so. I came up with a mathematical formula to calculate the actual size of the snake when I was told of an aquatic monster terrorizing some oxbow or lagoon. I took the length that I was quoted and divided it by the coefficient of fear and subtracted the constant of amazement of the person. This simple calculation often brought the number to something that made biological sense! By evaluating how scared the person was of snakes, I could guess the truth of the story.

In actuality, I developed a more scientific way to discard spurious reports of giant snakes. I asked the informant for three dimensions of the snake. On top of the length that people would normally volunteer, I requested the size of the snake's head and the width of its body. It is unlikely for a person who has not seen an actual giant to guess all dimensions within the proper ballpark. The reconstruction of the reported snake by these measures often rendered an animal with an extremely small head or as skinny as a noodle, which suggested to me that there was no such snake. There were, however, a few times when the measurements were internally consistent and I could not rule out that the person had, in fact, seen an extremely large snake.

Perhaps the most remarkable of such cases was in a report from the Beni in Bolivia. Reportedly, they have seen, *and captured*, a snake that was 17 meters long (56 feet) and 3,000 kilograms (6,600 lbs) with a head 70 centimeters (2.3 feet) long. This was the first report of such a large snake, and I would simply discard it as fiction if it were not for the fact that this story was internally consistent. True to my method, I needed to see if the measurements made sense. When I consulted my regressions, I calculated that a 17-meter-long snake would have a head 50 centimeters (1.6 feet) long and would weigh close to 2,500 kilograms (5,511 lbs). Five hundred kilograms (1,100 lbs) seems like a big difference in mass but it isn't, considering that the mass increases to the third power of the length. At this size range, a difference in length of half a meter yields a mass difference of several hundred kilograms. Thus, the 500-kilogram differential was really not a reason to discard the report.

When the report of this snake surfaced, I was working full time for *National Geographic* (NG) as a Field On-air Correspondent. Clearly, the NG executives were thrilled about finding such a big snake and the possibility of filming it. I spoke to them about my skepticism regarding this snake, but I also told them how this report passed my test to rule out nonsense. I explained to them that I didn't think this snake was quite this large, but even if it was 10 meters (33 feet), it would still be a formidable giant. In fact, if it was only 6 meters (20 feet) it still would be larger than any I had caught, and since the report said they had it in a fenced-in pond, we had nothing to lose. We could surely film a very large snake that I would be interested in wiring and following to study its behavior in the wild. NG had plans to buy all the land surrounding the area to protect this animal and all the habitat where it lives, contributing to the greater conservation effort—which is a pretty decent plan when you look at it.

The day before I was scheduled to leave for Bolivia I managed to make contact with the owner of the land where this snake lived. He told me that they did not have the snake fenced in, but it was easily found in a bend in the river of his property. This was a substantial change in the story, as it was not necessarily a secure deal. I heard many times of a snake that "certainly is there" only to find that there was no such snake. He also told me that his workers reported this snake to be 60 meters long (197 feet). This news was disconcerting because 60 meters was a lot harder to believe and it threw off the calculations that made me consider it. He explained to me that he believed his workers were greatly exaggerating the size of this snake; his guess was that it was only around 30 meters long (98 feet). *Only 30 meters!*

You can imagine how I felt about this news. Not wanting to give up, I pressed him to tell me as much as he could, trying to get an idea of the true size of the animal and the odds of finding it. Because of the size ambiguity, I asked him how close they had gotten to the snake. He told me they had seen it on the other side of the river about 600 meters (656 yards) away and had estimated its size based on the wake it produces when it swims.

With this news, I was downright depressed, because if I know something about anacondas I know they are very smooth swimmers. A 4.5-meter-long anaconda (15 feet) can glide effortlessly beneath water that barely covers it without producing so much as a ripple. It was impossible that these people could see a ripple, much less a wake, 600 meters away. By now I was sitting down, wondering if there was any point in continuing to look into this story. I asked if anybody had gotten closer than that. He said no one could not get very close because this was a very dangerous beast. He said a friend of his who runs a tourist operation in the area had gone at night and had been able to spot it from 120 meters (394 feet) away; the friend said you could see her glowing eyes, the size of baseballs, shining 30 centimeters (12 inches) apart from one another. Indicating how terrible this behemoth was, he said, "When this beast bellows the ground shakes all around you."

With this information, I realized what was going on. I have never heard an anaconda producing any sound that resembles a bellow. Even if I was willing to believe that the breathing sound I have heard an anaconda make when it is doing physical effort could sound like a bellow if the snake is 30 or 60 meters long, I simply couldn't buy the shining eyes. The reason some animals' eyes shine at night under a flashlight is because a structure on their retina, the *tapetum lucidum*, reflects the light that enters the pupils. This is an adaptation to magnify the light inside the eye. Some of the light escapes back out through the pupil, which is the eye-shine we see. This is rather common in nocturnal animals but very uncommon in snakes. Only one genus of snake, *Corallus*, is known to have it; anacondas do not have it. Even if I was willing to go along with all the other evidence of nonsense, with this one I had to give up: The physiology and structure of the eye would not change with a large snake. But at least now I knew what I was dealing with.

I called the NG chief executives and told them I would be happy to go to Bolivia to wrangle a large black caiman (*Melanosuchus niger*) for them to film, but they should not expect the reported beast to be anything that resembles an anaconda. A large aquatic animal that produces scary deep bellows, with a head 70 centimeters (27.6 inches) long and glowing eyes 30 centimeters (12 inches) apart, could easily be a 5- or 6-meter-long black caiman. They are known to inhabit the area. They were driven to near extinction for the skin trade and became scarce enough that most people have never seen them and don't know that they exist or that they can grow that large.

We never went to look for this behemoth, as all the evidence indicated that it was not a snake. Similarly, it is more than likely that the bellowing monster described in the introduction to this chapter was an Orinoco crocodile (*Crocodylus intermedius*), which shares these habitats with anacondas in the llanos. Federmann's soldiers could have seen one and heard the other and, not knowing much about either, made their best guess.

I could write a whole volume about the phony stories I've heard. I referred to this one because of how close we were to being fooled by it and its initial internal consistency. I am regularly contacted with stories about gargantuan snakes, and it is impossible to avoid the attraction that the story of an extremely large snake has. Sure, it would be cool to find a very large monster, but I'm a lot less interested in finding it than in learning about it. What is the life of a very large animal like? How does it differ from the smaller, but still very large, anacondas? What are the energy requirements of such big snakes? Why would they grow so large? Why not larger? The anacondas I have caught can go for a whole year without eating a meal and can attack a prey larger than their own mass. They drop smaller preys that are not worth their effort. Can a very large snake grow so much that no prey would be worth its effort?

Of the many questions that go through my mind, the ones I will try to address in this chapter focus on the reasons and limitations for large size in females. Why would an animal grow to such size? Why not larger? Does the blueprint of the snake body sets some limits on how much they can grow? We know that animals like, say, insects, are limited in how much they can grow due to the system of oxygen delivery to their tissues. If the insect is too bulky, diffusion of gases to its tissues is not quick enough for its oxygen delivery system to allow it to efficiently perform its metabolic functions. Is there a similar limit on snake size? Is there something inherent in the biology of snakes that prevents them from growing beyond a certain size? What is it? What is that size?

In general terms it is easy to point out advantages and disadvantages of large size. Being large has a number of benefits and some disadvantages. Large animals have fewer, or no, predators and can consume larger prey, and large females can have more, and larger, babies. On the other hand, being too large may cause some problems. For a predator, being too large might make it too conspicuous for its prey, and being too large leads to a very high cost of locomotion and high metabolic expenses associated with large size. Depending on how these and other opposing forces interact with each other, we would expect natural selection to select an efficient size. Also because of the extraordinary sexual size dimorphism, we need to consider selection forces acting on males and females differently. Considering how large females are, nearly five times the size of males, it is obvious that the evolutionary pressures for large size act more strongly on females than on males. In this chapter, I will deal with selection for large size in females and in a later one I will discuss selection pressure affecting size in males.

Reproductive Tradeoffs

One aspect in which natural selection definitely favors large size in females has to do with the reproductive output. The larger a female is, the more babies can develop in her body and the larger the reproductive output. Reproductive value or lifetime reproductive success is the number of potential offspring that an individual can leave in the population over its lifetime.[40] Natural selection should maximize it since it is a true measure of fitness.[41,42] In animals with undetermined growth, clutch size increases with female size due to an increase in the coelomic cavity (the size of the chamber the snake has inside her body, where her organs are); basically, a larger female has more room in her body to develop babies. So, the larger she gets, the more babies she can produce.

There are costs animals must face when they make reproductive decisions. If a female breeds, there are costs associated with breeding. Some of the costs invested in reproduction depend on fecundity (the number of babies that she has) and some of them do not. For instance, the energy invested in reproduction is dependent on fecundity because the female will need to expend more energy to produce more babies. One the other hand, there are costs that are independent of fecundity. For instance, the risk of being preyed on during mating or pregnancy and the odds of having parturition complications are not dependent on how many babies the female has. These two kinds of risks are important in determining what reproductive path natural selection takes. The fecundity-independent cost of reproduction (risk of disease, predation, or complications at birth) stay relatively stable through the life of the individual, so I will focus on the cost that varies with number of babies she produces.

To Breed or Not to Breed

A young adult female that has just reached maturity is under two opposing pressures: to breed right away and secure a few babies into the next generation or to skip reproduction, grow larger, and make more babies in a later year.

If she breeds, she will run the risks associated with reproduction, such as predation during mating and a long-term pregnancy during which she fasts. Fasting leaves her very thin and weak. Using her fat reserves for reproduction, she would not have much energy left for growth and would take longer to reach a size at which she has outgrown predators and can make a truly large reproductive investment. The next breeding season she will be only a little bit bigger and will not be able to produce many more babies than the previous time. So, her lifetime reproductive success will depend on many smaller reproductive events. The good thing is that she would have secured some offspring, which is not a bad idea if the reproduction-independent chances of death (predation, diseases, complications) are high.

On the other hand, the female who skips reproduction will use all her energy surplus to become larger, and she will be able to make many more babies the next year. If the female can increase significantly the number of offspring by increasing her size for 1 year, selection should favor skipping reproduction that year. Of course, a female that took this route would risk falling victim to predation or dying trying to get a meal, and she would not leave any descendants. She will not make the few babies she could have when she was small, nor will she produce the larger number of babies that a large female can make because she never became one.

The strategy that natural selection will favor depends on the snake's habitat, resources available, and the fertility-independent cost of reproduction, such as the odds of dying from one year to the next when she is not breeding. If the female has few resources and a high risk of predation, natural selection will favor her securing some babies as soon as she is big enough to breed. That way, if she were to be preyed on, she will, at least, have left a few babies for the next generation. So whether she breeds or not depends on the odds of surviving to a next year and the increase in size (and associated increase in fertility) she can attain during that year.[39,43]

It is not that different from the choices young adults face regarding their education. One can get a job right out of high school in the retail industry that pays lower wages but allows the person to become financially independent, or one can go to school, spend many years studying, and come out with, say, a law degree that will grant one a higher income. A young adult may choose one or the other path depending on the circumstances. The person who starts making money early in life and does not invest in an education will probably not obtain very high wages in his or her lifetime but will start an adult life sooner (including reproduction). The person who continues in college for longer has a better chance of making higher wages after graduation but will have to delay financial independence (and often reproduction) for a few years. Of course, the parallel goes only so far. Humans make conscious decisions while weighing their choices. Non-human animals have likely other ways to make these decisions, but the results can be interpreted in a similar way.

Another element that plays a role in how the animal invests its resources is the future reproductive success, which is the reproductive value that the animal has left at any given time of its life. Young animals that have a long life span and can produce many clutches in the future should invest proportionally less per individual reproduction because all energy used in offspring is at the expense of future reproduction. On the other hand, older animals, who have a shorter time left and a lower potential for future reproduction, are expected to invest more in reproduction as they do not forfeit a lot of future reproduction by making a very strong reproductive investment. This tradeoff can be exemplified by the unfortunately familiar case of a teenaged single mother who gives her baby to her parents to raise. Since she is so young, she has a lot of future reproductive potential, so she invests little in that child and focuses on her own life (studying, starting a career) so that she can get a leg up in her life. On the other hand, her parents, who already have passed their reproductive life, have little to lose by investing resources raising a grandchild. In fact, since the grandchild has their genes, they actually gain inclusive fitness by doing it.[44]

I had the opportunity to see this theoretical paradigm in practice with a loud bang of snapping jaws and sharp teeth. I was helping my friend and mentor John Thorbjarnarson in his collection of data on the reproductive biology of the spectacled caiman (*Caiman crocodilus*) when he was doing his dissertation. We located the females' nests and opened them up to count and measure the eggs present in the nest. Female spectacled caimans guard their nest from predators and will attack intruders that come near it. However, their defense is normally directed to the more regular nest predators such as tegus (*Tupinambis teguixin*), opossums (*Didelphis marsupialis*), raccoons (*Procyon cancrivorous*), and so forth. These are smaller egg predators (between 500 grams and 5 kilograms) and cannot do any harm to an adult caiman. In fact, these egg predators can easily turn into a nice meal for a guarding female. So, the guarding females did not attack us when we went to count and measure their eggs. Two adult humans are far larger than the regular nest predators the guarding females have to deal with. Whenever we got to a nest, we saw the females nearby, and sometimes they would bellow or make noise to show their presence, but since we were undeterred the females just kept a safe distance from us. We always brought along poles and snares just in case we needed to keep a female at bay or tie her up, but we did not have to use them.

There was an exception to this, though. One of the females that was guarding the nest was very skinny. She seemed to be victim of a wasting syndrome that we see on occasion among caimans in the llanos. The animals become progressively very thin. They can live for a long time with this condition, and it may be associated with some viral infection. It is likely something similar to the wasting syndrome that can be seen in other wildlife in North America. This female had just that appearance. As we approached the nest, she charged at us decidedly. We were forced to use our longer legs and superior terrestrial ground locomotion skills to escape the snapping jaws of the cachexic caiman. It was a stark contrast with the behavior of other healthier, stronger-looking females, who did not dare to attack us to protect their nests. It was very surprising to see this dying animal muster the courage and strength to come after us with such resolve.

Future reproductive success was exactly what was in play here. The healthier females had lots of other nesting seasons in the future to protect. If they died trying to protect this one clutch, they would be risking all future clutches, so the best investment was to give up on a clutch that is being attacked by two large humans instead of risking their life (and future clutches) in a daring defense. On the other hand, the one with the wasting syndrome was probably going to die soon of the disease, so she had nothing to lose by fighting to the death defending her last clutch.

Other Decisions: Large or Small Babies?

A female that is going to breed faces another question: How should she invest her breeding resources? She can produce a large number of neonates of small size or a few offspring of very large size. Large offspring may reach a larger size more quickly and should suffer less predation. Larger individuals can also kill and subdue larger and more diverse prey than small ones, an additional benefit of having larger offspring size.[45,46] On the other hand, it would also benefit a female to have as many offspring as possible, certainly if the risk of predation is high. Since the amount of energy available for reproduction is limited, there's a tradeoff between the size of the neonates and the number of them in every clutch.[47,48] Generally, if the animals have a high mortality early in life it would benefit the mother to have many offspring (r-selected strategy); on the other hand, if the environment is relatively secure it is best for her to have few but very well-endowed offspring (K-selected strategy).[49, 50] Thus, natural selection should optimize clutch size in order to have the largest number of offspring that have a good chance of survival.[43,51]

So the theory predicts changes in reproductive investment. Individuals that have many more clutches to lay in the future are expected to have a smaller reproductive investment per clutch as a consequence of their small size and the fact that they need to divide their energy surplus into growth and reproduction. Older (larger) individuals, on the other hand, have less to lose by making a big reproductive investment, since they do not forfeit much growth by breeding, and they need fewer resources for the future as their expected future is shorter. They are expected to make larger reproductive investments.

I am not saying that the animal makes a conscious decision about how to invest her resources. Neither was the cachexic caiman doing some advanced evolutionary mental calculation when she decided to attack us. All these decisions are shaped by natural selection and have some proximate causation, often associated with some private experience, like anger, fear, or some other emotion, and they are often triggered by some physiological variable very different from the evolutionary level of analysis we have been discussing.

Natural selection has shaped these behaviors *as if* the animals were doing mental evolutionary calculations. This is the same reason grandparents normally are so devoted to their grandchildren. Natural selection shapes these variables, increasing the reproductive outcome of individuals in the following generations. Thus, the proximate reasons (the protective feelings of the caiman) and the evolutionary explanation of the adaptive behavior are not at odds, or mutually exclusive; rather, they are simply addressing different levels of analysis. Neither does an individual need to understand the evolutionary reason in order to act on the proximate, emotional, triggers of the behaviors.

Levels of Analysis

When trying to explain biological phenomena, there are often alternative explanations that are not mutually exclusive; rather, they simply deal with different levels of analysis.[52] I will not attempt a thorough review of the issue here but will provide a mundane example. Let's say that a person enters a dark room and flicks the light switch, and the bulb starts radiating bright light. Why did the light go on? There are several explanations. One is that the person desired more light in the room and turned on the light. Another one is that flicking the switch closed the circuit and the bulb became energized. There is yet another explanation involving the temperature of the element in the bulb (assuming it was incandescent): Since the element in the bulb has a high resistance to electricity, it heats up when electricity goes through it. Because all elements emit radiation in a frequency proportional to the fourth power of their absolute temperature, the resistance started producing visible light because of the heat produced by the electricity going through it. We can offer many other explanations for the phenomenon; perhaps as a child the person had an awful experience in the dark.[53]

Which is the correct explanation? There is no single correct answer. They are all correct answers, but they are addressing different levels of analysis. In these situations, we often find several alternative explanations that are not mutually exclusive; they simply answer the questions at different levels.

For instance, when the cachexic female attacked us, she was probably reacting to a deep feeling of attachment to her clutch. However, this can also be explained evolutionarily, as I have mentioned. As we move forward trying to understand anaconda reproduction it is important to keep this in mind.

Many of the explanations we will be seeking are at the evolutionary level, dealing with natural selection shaping behavior and the morphology and physiology of the animals, but the animals may be responding to different proximate causations or triggers.

Data Collection

To disentangle the secrets of the anaconda's mating system, I gathered pregnant females (Figure 6.1) from bodies of water and surrounding areas, mostly during the dry season but also during the wet season.[4] Of course, since females are larger to start with and pregnant females are even bulkier, collecting these data involved quite a fair amount of snake wrangling. Although I was thrilled with the idea, it seemed quite daunting to me at the beginning of the project, certainly before I had learned the tricks of the trade. I can't count how many times someone has asked me if I'm afraid to catch a big anaconda. Well, am I?

Hanging on to Your Fears

Ever since I was in college, I have been attracted to dangerous activities. While I studied biology at Universidad Central de Venezuela, I served as a fireman at the university station and sought any kind of adrenaline-producing activity, including scuba diving, search-and-rescue (SAR) operations, rock climbing, and pretty much anything that came to mind. In my spare time as a fireman, I made a sport of climbing the external surface of all the buildings on campus (an activity for which some people have gained significant reputations lately in the United States!) in the interest of practicing SAR techniques and, truth be told, seeking adrenaline highs. Few of my friends were surprised to see me pursue a research career that involved handling dangerous reptiles.

I want to spend a few lines addressing this issue because one thing I learned while working as a firefighter was that your fears are your best friends when you are in a situation that might be dangerous. I learned this the hard way. I will mention briefly an accident I had during a training maneuver—due to the knowledge I gained from it, not because I feel very proud of it. It is not uncommon to hear people who work with snakes boasting of accidents they have had, like a snake bite, without seeming to realize that when you are bitten by a snake in the process of handling it or when you have a fall while doing a SAR maneuver, it is almost invariably because you made a mistake. You don't hear surgeons boasting about leaving scissors inside the patient after a surgery. You don't hear taxi drivers bragging about totaling their cab and killing their passengers. You don't hear accountants bragging about miscalculating someone's taxes and landing him in jail. These are all mistakes that happen on occasion, but when they occur, the responsible party is often sorry about them or even embarrassed. As a snake-handling friend of mine once told me regarding a bite he had received, "When you land on your pitchfork, you don't brag about it." Yet all too often I hear people referring to accidents they had handling dangerous reptiles as if they were medals of honor!

At the risk of falling into that group, I will refer to an accident I had while preparing an SAR maneuver at the fire department because it really helps me address the issue of fear. We were preparing a demonstration for the following day that was going to take place in the university library at Universidad Central de Venezuela, a 12-story building. I was supposed to use a rope as a bridge to move from the top of the building to the ground, which required me to have a safety carabiner attached to my chest harness. For practice, we were doing this on the fire department's regular practice grounds, from a ledge on a two-story building. Since this was a demonstration, and I was the SAR instructor, I was expected to give a pretty flawless, and even showy, performance. During my first rehearsal, though, I fell off the rope in the process of climbing on it, which was not unusual since it was quite a tricky move. When I was dislodged from the rope, I didn't even try to hang on to it, fearing that the wild shaking of the rope would injure my shoulder joint. I just let it go and waited for the carabiner to hold me. Nothing happened because I had the safety carabiner attached to my harness, so I could simply climb back on the rope and finish the maneuver.

However, since this was going to be a public demonstration, I wanted to do it without a glitch. Intent on getting it right the second time, I repeated the maneuver but forgot to secure the carabiner that had prevented me from falling to the ground the first time. On the second try, the rope (which was more elastic than it should have been) again bounced wildly when I tried to climb on it, just like the first time, and just like the first time I was dislodged from the top of it. I again let go of the rope to protect my shoulder joint, sure that the carabiner would catch me again. I recall vividly my surprise when

I saw the rope going up and up and up as I fell to the ground. I managed to land on all fours, which spared my spine from damage. I recall the loud bang of my helmet on the pavement of the parking lot before I lost consciousness. The helmet cracked, absorbing the blow that was intended for my skull. I broke all four of my limbs and was in a wheelchair for the following few months with plenty of time to think about it.

Many of my colleagues at the fire department were surprised at my fall, since I had worked and trained on that ledge so often that it was like my playground. And that was precisely the problem: The ledge was all too familiar to me. I had used it for years, for descending, ascending, and all kinds of maneuvers. I had taught countless firemen apprentices how to rappel from it. I had told them countless times not to be afraid, and I really felt completely safe on it. The maneuver I was preparing for was from a 12-story building. That distracted me, and when I was feeling very comfortable on my ledge, I forgot the minimum bit of fear that one must have when one engages in something that is dangerous.

What I learned is that fear is not such a bad thing. Of course, too much fear can paralyze you, and that might not necessarily be good, but having a healthy amount of fear when you deal with something dangerous is quite a good idea. Nothing but your fear will ensure that you hook up your carabiner. Only fear will remind you to double-check your safety protocol, and only fear will drive to you take the extra steps needed for your security. It is no different when you are climbing a mountain, handling a dangerous animal, or even diffusing a bomb (for those who do that kind of stuff). If you engage in dangerous activities, you must treasure that fear, because it's the only thing that stands between you and disaster.

On occasion you'll hear about a person who does something dangerous (say, handling dangerous animals, fighting a fire, or serving as a Navy SEAL) and claims not to have any fear. In most cases it just isn't true. Often the person is just bragging about having no fear, wrongly believing that fear is something bad or something to be ashamed of. If someone is truly unafraid, it's only a matter of time before an accident will happen. My most sincere advice for anyone doing something dangerous is that if you ever realize you have no fear of what you are doing, seek it! Do all you can to retrieve some of that healthy fear, because it's your greatest ally when danger is around you. Think of what would happen if you made a mistake. Try to envision the worst that could happen. Only a healthy dose of fear will preserve your life, and you must be grateful to have it.

Collecting Breeding Females

Pregnant females bask frequently and seek higher elevations in the wet season, the few places that are not flooded in the flat llanos landscape. This gave me a prime opportunity to find pregnant females, because easily accessible places like roads or riverbanks were great places for looking for snakes (see Figure 6.1). However, dealing with the largest animals proved to be the most adventurous part of the study. I recall distinctly the time we caught Jane in Hato El Frío. Alejandro Arranz was the manager of the tourist station at El Frio and a very enthusiastic snake lover. He was originally from Spain and had been working in the tourist station for a few years, so he knew a lot about the natural history of the animals (Figure 6.2). He gave us support and help, including going with us in a 4-meter-long (14-foot-long) aluminum boat looking for snakes on the riverbanks. The day we caught Jane, there were four people in the boat: myself, María Muñoz, Alejandro leading the boat, and Peter Strimple, a friend and snake lover who had come to learn about anacondas and give us a hand (one of the people Bill contacted to come help). We went to Caño Macanillal where, although everything was flooded, there were trees half-covered by water that provided a good place for the anacondas to bask on their branches. We found a number of them basking, but it was not easy to get close without disturbing them. They could easily slide down into the water and disappear from view as the water was at least 7 meters (21 feet) deep.

We decided to approach a snake that, given its bulk, seemed to be worth our effort. On my command, the three men reached overboard to grab the snake while María, on the prow, threw all of her weight (53 kg [115 lbs]) to the other side of the boat to prevent it from capsizing. To our surprise we ended up with two tails in our hands, which explained why "the" snake looked so much larger than the others. Alejandro weighed about 60 kilograms (143 lbs), while Peter was some 105 kilograms (230 lbs). Noticing that one snake was substantially larger than the other, I gave the larger tail to Peter to hang on to so I could team up with Alejandro to pull the smaller one into the boat first. This would free the three of us in our efforts to subdue the bigger one later. Of course, "smaller" was a relative term, since the snake that Alejandro and I had was nearly 30 centimeters (1 foot) across and our hands were hardly big enough to get a good grip on the tail, even the lower parts of it.

I saw with disappointment that Peter's snake slipped out of his grasp before we had a chance to help or decide which snake to struggle with. I tried to reach the lost tail to no avail. I gave poor Peter an earful for letting go of the snake without asking for help, but it was too late. We had to resign ourselves to the smaller animal. We all focused on the tail of the snake we had left, the one we later named Jane. But Jane got a good grip on the keel of the boat and was pulling so hard that slowly but surely, inch by inch, scale by scale, she was slipping from our hands without our being able to hang on to her. María was doing acrobatics on the prow by leaning as far as she could to keep the boat from capsizing but wasn't having much success, as the three of us far outweighed her and we were leaning heavily over the other side. The farther we tried to reach over to pull Jane in, the bigger the risk that the boat might tip over. Not wanting to give up on this one, I got out of the boat and held on to the other side of the edge of the boat so I could push with my feet on the snake's body under the boat. I was trying to loosen her grip on the keel so Alejandro and Peter could pull her in. As I pushed harder and harder to release her from the keel, she got annoyed and bit my ankle.

It was one of those moments in which a million things passed through my mind in a split second. I felt a chill travel all over my body at the realization that the snake was stronger than all of us together. If she pulled me down, my arms would be a poor match for her strength. My friends on the boat would not be of much help either, as it was painfully obvious that the snake was stronger than our combined strength. Jane was secured under the water and the water was too deep and murky for anyone to go under and help me if she succeeded in dragging me down. The water was too deep for any human to put up a decent fight against a giant aquatic snake. Fortunately, I breathed with relief when I felt her let go of my ankle. Without thinking, due to the adrenaline rush, and intent on not losing the snake, I continued pushing even harder to loosen her from her keel. This time she bit me again and held on to my foot.

I felt that chill again and held still, not moving my foot and wondering what was next. I don't know how long she held on this time, but it felt like an eternity. I was powerless and at the mercy of the snake, waiting for the coils to start climbing around my legs and dragging me from the edge of the boat under the water. However, it was probably only a few seconds before she let go of my foot again.

Realizing that it was not such a good idea to try my luck at another attempt with the same maneuver, I tried something else. I climbed halfway into the boat and balanced myself on my stomach over the edge of the boat, asking a very surprised Peter to hand me a piece of rope that was on the bottom of the boat. I went down again, this time holding my breath in the murky water. Since I couldn't see anything, I ran my hand along the surface of the bottom of the boat toward the keel until I felt the massive scaly body of the snake. I was feeling blindly with great anxiety since I didn't know what to expect if the snake grabbed me again on the arms or torso. I made sure I had located a thick enough part of the snake and summoned up my memories of serving in the fire department.

When I taught SAR training, I always insisted that my students had to be able to tie all the knots in complete darkness because you never knew in which conditions you'll need to make a knot when you're out there on a rescue mission; smoke or darkness might prevent you from seeing what you're doing. I had no idea how right I was! Blindly, I made a clove hitch around the snake's body, making sure to hold the rope far away from the snake when I tightened it so that, in case she snapped at the point where she felt the pressure of the rope, she wouldn't find my hands. I swam to the surface triumphant with one end of the rope in my hand, feeling that the worst part was behind me. I told Alejandro and Peter to let go of the snake's tail and instead pull up the rope. I will never forget the puzzled look on their faces. They never knew what I was doing and were mystified at my command to let go of the snake that they had made so much effort to hang on to. I climbed back on the boat, and since the snake's grasp was meant to hold from one side of the boat, pulling her from the other side was easy and we met with no resistance. Before we knew it, we had a very large and very ornery snake thrashing around and snapping at everything (including Peter's shoe) on the bottom of the small boat—but it was now on our terms. We wrestled with her and eventually controlled her. Jane (named after Peter's wife) was 4.86 meters (16 feet) long and very pregnant. I still wonder what would have happened if we had tried to catch the bigger snake!

Jane was but one of the contributors to my study. I gathered 50 females over the years and had them in outdoor enclosures with pits holding water where they could hide as well as shady and sunny spots where they could thermoregulate (Figure 6.3). The first year I kept live chickens in the enclosure to provide food for the basking females, but none of them ate any during pregnancy. At times I felt more like a chicken farmer than a snake researcher! The fact that anacondas do not feed during pregnancy is rather surprising for anyone familiar with the good appetite a pregnant woman has during her pregnancy, and perhaps is a consequence of the high risk anacondas face while taking a meal (see Chapter 5).

An Important Decision

Catching and holding large breeding females taught me just how strong these animals can be and how motivated they can be to breed when their time comes. I had caught a female who was fairly large but by no means exceptionally large. She was about 4.3 meters (14 feet) long, in great shape and very strong. Since I caught her in the afternoon, I did not have time to process her and release her that day. I put her in a metal drum overnight to process her the following day. Because I knew the strength of the snakes, I put two cinderblocks on top of the lid of the drum to prevent her from getting out. At this time my dog Chuka was already a well-established member of my household. She came to the field all the time with me and had a bed in the living room. She knew not to come to my bedroom because she was often muddy. However, that night, Chuka came to my bedroom, pushed the door open, and woke me up by licking my face. I was surprised by her behavior and took her to her bed, in the dark living room, and instructed her to lie down. She did. I went back to bed, only to be awakened again as I was falling sleep by a very anxious Chuka. I shoved her out of the room one more time, but before I was back in my hammock, Chuka had pushed the door open one more time. She looked scared; her stumpy tail was tucked in and she was acting very oddly. That day we had seen a puma not far from the station, and I wondered if she had heard something or smelled it nearby. I went out to the living room and waited quietly in the dark, hoping I could hear something or figure out why she was acting so strangely.

It might have been 5 minutes later when I heard a noise, inside the house! A living room chair had moved a few inches. Chuka was sitting next to me, so I knew it wasn't her. We were not alone!

Turning on the light, I saw the anaconda with its full 4.3 meters stretched out in the living room looking for a way out of the locked-up house. Poor Chuka had to face the dilemma of being scolded by me if she got into my bedroom or spending time alone with a giant snake loose in the living room! I woke up the rest of the crew and we soon had Arnolda (as we named her for her strength) back in the bucket. She had lifted the two cinderblocks we had placed on the lid to get out. I placed two more cinderblocks and a bucket of water, which I deemed would do the job. However, 30 minutes later, Chuka was back, licking my face to wake me up. This time I went straight to the drum and found Arnolda halfway out, having lifted all the weight I placed on the lid. Arnolda was in reproductive condition when I caught her and probably felt a very strong urge to be out in the swamp, since the mating season is not very long.

As the end of my time for living in the field drew near, I wondered with concern what to do with Chuka. She had grown up in the llanos as a feral dog but by now had spent more than a year with me as my partner. The ranch managers did not want feral dogs around because they prey on capybaras and other animals that tourists pay good money to see, so Chuka was not really welcome on the ranch—certainly not now that she was not skittish, had lost all fear of people, and was actually very friendly and loving with just about anybody. I thought of taking Chuka to Caracas to live with my mother. She always loved dogs but never had one of her own. The one dog that I had before I imposed on her since I was living in her house. She loved it and enjoyed every bit of the time she was with us. But who was I to decide Chuka's life? I didn't view her as "my" dog. I was not her master; rather, we were friends who had decided to spend time together. That I had the power to decide her future was not a foregone conclusion, in my mind.

Yet leaving Chuka on her own would probably mean that she would be shot by the ranch managers. They did not want feral dogs and often shot them on sight. Chuka was now tame and did not have the "edge" that kept other feral dogs alive by avoiding humans. Leaving her probably would have meant a very short life for her. Even if I managed to convince the ranchers to leave her alone or if she found her old pack, the life of a feral dog is not a very long one. Predators abound in the llanos; there are parasites and shortages of food. I could make sure she would have a long and pampered life if I took her to be a pet dog as opposed to a feral one. Yet, who was I to decide? I knew that if I had the choice, I would never leave a life of freedom in the llanos chasing critters and running around, short as it may be, for a long life confined in an apartment, or even with a big yard. Why would I impose that on Chuka?

As I was debating this issue, I went to the field to track radiotagged snakes. On my way I had to pass a river that this time was a bit deeper than usual. It was nothing my horse couldn't ford but was far deeper for Chuka. In these situations, she normally stood back and, after looking up and down for a place to cross, she would go back to my house and wait for me to return in the evening. But this day was different. Chuka was intent on following me all the way, as I moved away from the river on the other bank she barked and barked on her side of the river. Then I heard a plunge in the water and the barking stopped. I looked back, wondering if she was OK, only to see her braving the current, swimming with decided strokes in my direction! She got to my side and continued following me in the swamp, with the water up to her chest most of the time and often having to swim to keep up. I kept an eye on her to make sure that she was OK because of the many caimans and anacondas in the area where we were, but I read in her behavior the answer to my questions.

Chuka was terrified of the water. She knew all too well what lurked in those rivers: anacondas, caimans, crocodiles, electric eels, and piranhas, just to mention a few. She always avoided water she could not cross with a few long, powerful strides. Up until then she never dared go in deep water, let alone swim over a river and follow me for the whole day in deep-ish water. What I read in her actions was that she was ready to follow me no matter what. She was ready to brave all the scary predators of the river so long as she could follow me. This is what made me decide to take her with me. And by "with me," I didn't mean to live in an apartment with my mother in Caracas; no, I took her home with me to the United States. I had to do some paperwork to "import" her into the country but surprisingly less than I feared. Chuka enjoyed a long life by my side. I always tried to give her access to wild areas where she could roam far and wide, the way she grew up in the llanos. On more than a few occasions she came back with her breath smelling like a butcher shop after having caught a squirrel or small rabbit out in the bush. I never repented my decision to bring her home. She passed quietly at the ripe old age of 17 years, well past the life expectancy for a dog her size. Since she was 12, she had amazed veterinarians, who always surprised by her good health. "She has better blood chemistry than most dogs half her age," they would declare on getting her results. There's something to be said about the results of natural selection on genetic fitness!

A Pregnant Figure

Most of us have gone through the experience of congratulating a woman, believing her to be pregnant, only to find out with embarrassment that she is not. I know I have. However, the more I studied anacondas the more confident I became that this is not a concern when dealing with female anacondas. And I don't mean that female anacondas are not self-conscious about their weight; in fact, males are attracted to fatter females (see Chapter 8). I'm referring to the fact that it is very easy to reliable determine if a female is pregnant just by looking at the shape of her body. Early in the study I realized that some females were substantially fatter than others. This is true especially among those found in breeding aggregations. I realized that I could pretty much predict if a female was going to breed by looking at how fat she was. I decided to develop an index of obesity, based on the length and the mass, in order to measure objectively the condition of the animals.

There are methods to study condition of the animals that look at the body measurements of the animals and come up with an index of condition. A popular index is the residuals of the log/log regression of mass and length. However, this method has a weakness: If the average animal's body becomes thinner (or fatter) as it grows larger, the residuals will be calculated with the new value for that size. In other words, the standard with which we measure obesity may change with the average obesity of the animal. This is the exact equivalent of measuring the length of an animal with an elastic measuring tape!

I developed a condition index that was based on a simple transformation of the mass–length relation. It is unbiased and size independent, fulfilling the requirements of simplicity, appropriateness to the particular dataset, and statistical correctness. It is inspired by the equation of a cylinder (the closest geometrical model to a snake's body), which shows that volume (and thus mass) is a cubic function of length. It turns out that this index is mathematically identical to Fulton's index, a well-known index used in fisheries.[54] This index assumes that larger animals have the same shape as smaller ones (isometry). In other words, it assumes that there is not an ontogenetic change in shape other than those due to condition. Thus, I labeled it "condition with isometry assumption" (CIA).

I used data from 660 animals to analyze the mathematical appropriateness of the index. The relationship between mass and snout–vent length (SVL) of 660 animals is an exponential relationship (Figure 6.4). By calculating the cubic root of the mass, this relationship becomes linear (Figure 6.5). Finally, by dividing the mass's cubic root by the SVL, it is possible to obtain an unbiased and size-independent estimate of how fat the animal is that is independent of the animal's size (Figure 6.6). For the sake of representation, we can multiply this index by 10 in order to obtain an index that is easier to handle. The condition index proved to be a size-independent and unbiased tool to assess obesity in a large range of sizes, and it is normally distributed around the mean in a probability plot.

To test if this index was any good at assessing pregnancy, I needed a sample of the population that I knew for a fact was pregnant and another group

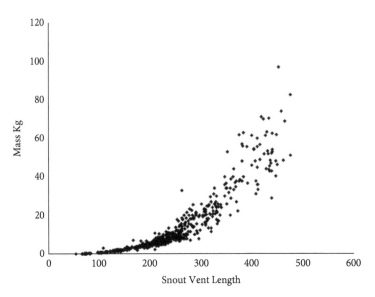

Figure 6.4 Relationship of mass and SVL in 660 wild-caught anacondas over a 6-year period.[15]

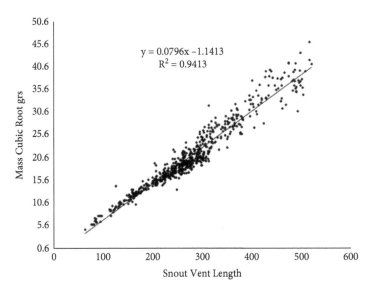

$$y = 0.0796x - 1.1413$$
$$R^2 = 0.9413$$

Figure 6.5 Relationship between the cubic root of the mass and the SVL in anacondas.[15]

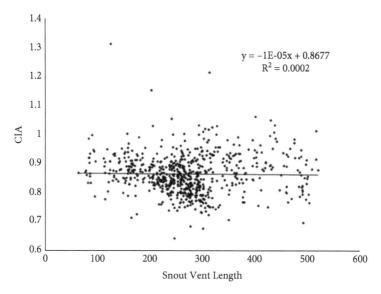

Figure 6.6 Condition index of the data from 660 anacondas from the wild population. The CIA proved to be normal in a normal probability plot.[15]

that I knew was not. Based on the bulk of the animal I could make a good guess, but I needed certainty about the breeding condition of the specimens. I resorted to an anthropomorphic point of view: What would a woman do to find out if she was pregnant? Well, an ultrasound is a very efficient measure to know if the female is pregnant or, in anacondas, if she has eggs developed ready for fertilization. Of course, in 1992 there was no fancy portable ultrasound equipment, and I certainly did not have the budget to buy one even if they had been developed by then. Instead, I contacted a local obstetrician/gynecologist in the town of San Fernando de Apure and asked him if he could help me.

A good thing about anacondas is that because they are so charismatic, there is always somebody interested in getting involved with them. The good doctor was curious to see if his ultrasound machine would work for anacondas as well and if he could recognize embryos or eggs in an anaconda with his knowledge of human pregnancy. He agreed to help me and asked me to bring the snake to his practice. Worried about not scaring his patients, I offered to come on a weekend or around his regular schedule. He did not seem worried about it. He asked me to come at 7 a.m., and since his patients normally arrived at 9 a.m. or later there should be no problem.

The procedure worked like a charm (Figure 6.7). We could see, count, and measure the embryos and eggs in females. Also, after seeing how eggs look in the ultrasound, the doctor felt confident that when we did not see them, it was because the females were not breeding.

Of course, sometimes our examination lasted a little longer, and by the time we came out of the ultrasound room, there were already some patients in the waiting room. It was rather awkward to roll a 208-liter (55-gallon) barrel containing a 4.6-meter-long (15-foot-long) snake through the waiting room. On occasions, some wide-eyed patients inquired what it was that we were dragging in that large barrel. My only answer was "Nothing really; we just work with the equipment here." This was technically accurate. I wanted them to think that I was a maintenance person who was hauling some equipment fluid from the machines of the doctor's practice. However, judging by their scared looks, I don't think it worked too well all the time!

Clearly, females need to gather energy to start reproduction, and because they do not feed during pregnancy, they need to gather all the energy they can before the next reproductive event to make sure they can make it safely through the process. It made sense that females who were about to breed had to be fatter than those that were not going to breed. These are called "capital breeders" (from the economic equivalent of someone who pays with capital up front for an investment) as opposed to "budget breeders" (someone who makes monthly installment payments, not having the initial capital for the purchase). So, having an accurate way of measuring obesity is a good way to understand pregnancy. Thanks to the ultrasounds and to some snakes I had observed either mating or giving birth, I had a number of snakes that I knew were pregnant and others that I knew were not pregnant. By graphing the condition of the animals whose reproductive status I knew, I found a clear cutoff that successfully predicted pregnancy. There was a significant difference between the mean CIA ($t = 12.1$; Figure 6.8) in pregnant (9.40) versus non-pregnant females (8.11). Note that animals with a CIA above 0.875 were most times breeding while animals below it were not. Note also that most observations (82 out of 87; 94.3%) were classified properly using this criterion. Four out of five of the mistakes were non-breeding females that were improperly classified as breeding. Even though I dropped from my analyses any animal that appeared to have had a recent meal, some animals may have had some partially digested prey that went undetected and that might have inflated their mass, resulting in an artificially higher condition than they really had. This CIA allowed me to predict pregnancy of females that were

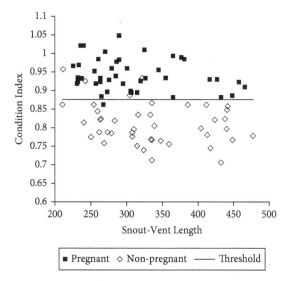

Figure 6.8 Condition index of breeding and non-breeding female anacondas. Notice that most non-breeding females had a CIA below 0.875, while most breeding females were above this cutoff.[15]

about to mate or that bred in a given year, which proved to be priceless in the development of the project.

Expecting Mothers

A lot of people are surprised to hear that anacondas bear live babies because most people believe that all reptiles are oviparous. Viviparity has appeared several times in the evolutionary history of squamates (lizards and snakes) and is not even uncommon. About half of the snakes in the world lay eggs and the other half bear live babies. It has been argued that viviparity is a strategy to deal with cold temperatures since the female can bask and warm up the eggs during the heat of the day and hide in a cave during the cold part of the day and night, keeping her eggs warmer than the average environmental temperature. However, I am unconvinced by this explanation. For starters, if feels like a temperocentric explanation, meaning that scientists from temperate regions (where most snake biologists are from) may overestimate the importance of cold weather. Even in the most northern latitudes where snakes occur, there are at least 45 to 60 days when the temperature is warm enough to incubate the eggs of snakes, so clutches laid in the right place (warm spots) should have enough heat to hatch. Furthermore, viviparity is quite common among tropical snakes and tropical clades living where the average temperature is 30°C (86°F), the target temperature to incubate the eggs. So even if cold temperatures might be the selection pressure for the evolution of viviparity in some clades of snakes, it might not be the only one (see later in the chapter for more information).

As I have argued, we can only ask the questions that our biases allow us to ask. The important thing is that we must be aware of our biases and how they might influence our views and interpretations. With my tropic-centric bias, I would think that things like fungi, ants, and other predators (common in the high diversity of the tropics) that may prey on the clutch would be a most important selection pressure for the mother to keep her babies protected inside her body, certainly if she is large enough and doesn't have much fear of predators. Even the unpredictable cycles of flooding, which are rather common in the Amazonian and Orinoco basin, may be a good explanation for the evolution of movable clutches that would not die if the water level rose too much. This may be the reason that neotropical boas are all viviparous while Old World pythons are egg layers.

Whatever the reason, anacondas do carry their eggs in their body through the whole term, which lasts approximately 7 months (from mating to birth). To study their pregnancy, I had some animals with transmitters that I followed in the wild. I also collected animals from the wild that I knew were going to give birth either because I saw them mating or because the CIA indicated they were going to breed that year. The telemetry follow-up of pregnant females told me a few important things. One was that pregnant females bask often next to the riverbanks or highlands, where they are easier to find (see Figure 6.1), but also that anacondas do not bask unless they have either a bad wound or are pregnant. This knowledge of their behavior, along with the CIA, let me gather pregnant females that I could catch before they gave birth and place them in captivity so I could get the babies; I wasn't sure I could do this from the animals in the wild. After birth I let the babies go in the same place where I caught the mother, or a nearby place with suitable habitat (Figure 6.9).

So, before the parturition time I got into the business of gathering females so that they could give birth in captivity. I went to Caño Macanillal with Alejandro. It was at the height of the wet season. Due to the flooding, we couldn't drive a car all the way to the river; instead, we had to travel by boat for quite a while through the flooded savanna. Since we knew that pregnant females seek high ground on the upper banks for basking, Alejandro scanned the shore slowly from the motorboat while I searched the riverbank on foot. At some point I realized I had been walking for some time without seeing Alejandro, and I backtracked to see what the delay was. I was surprised to see the boat, empty, with the engine running, spinning in circles in the middle of the river around a dead tree. I called and called for Alejandro with no answer until finally I heard him calling to me weakly over 500 meters (547 yards) downstream of the abandoned boat.

He told me from across the river (about 50 meters [164 feet] wide) that he had been attacked by killer bees and just barely managed to escape by jumping into the water after a lot of "fighting" with them. The wisest action might have been to motor away, but being surprised by the bees, he panicked and tried to swat the bees away. Fighting with them is, not surprisingly, unsuccessful, especially when there may be as many as 60,000 bees whose primary function is to defend their hive at all costs. When you start killing them, "defending yourself," you only convey to the bees the idea that you are indeed someone "who kills bees" and therefore a threat. Finally, Alejandro fell out of the boat and swam away while the loud engine noise kept the angry cloud of bees hovering over the boat as it spun in circles around their hive, which was in a hole in the dead tree in the river.

Africanized bees (all you find in the llanos) are very defensive, and if a person has an encounter with them, it is normally a serious one. At that point I didn't know how many bees had stung Alejandro, but I had reason to believe that the number was quite high. I knew he needed to be taken to the hospital ASAP. However, we were in a very remote area, and worst of all our boat (only means of transportation during the wet season) was spinning out of control in the middle of the river, escorted by hundreds of enraged bees. I also knew that the noise of the engine and the heat of the day at 1 p.m. made for a very inconvenient scenario to rescue the boat. The boat would stop making noise when it ran out of gas, but then we would be out of gas and stranded in the middle of nowhere. Alejandro was alone on the other shore with an undetermined number of stings, and there was no other hope of taking him to a hospital than my rescuing the boat.

I knew that in wet season the piranhas, although always present, would be at low density and unlikely to be a threat. I was not so confident, however, regarding the 3.5-meter-long (11.5-foot-long) Orinoco crocodile I had just seen 100 meters (328 feet) away, not to mention the many other crocs that I had helped reintroduce to the area as part of a conservation effort, and that were commonly seen in that stretch of river. Still, my major concern was the bees still buzzing over our boat. I yanked my straw hat down over my ears, secured my sunglasses, and swam to the boat, paddling doggy-style and trying to minimize my splashing of water on the surface. I made sure that my hat visor was too close to the water for the bees to fly in but open enough to let me breathe and see where I was going.

I approached the spinning boat, made a quick lunge over the backside, and hit the "off" button of the engine. I ducked down again to my safe position in the water and heard an endless number of bees hitting my hat and buzzing threateningly around my head. I resisted the temptation to fight them and thus expose my body out of the water, hoping desperately that they didn't find the opening that I was using to breathe. Eventually I grabbed the keel of the boat and slowly dragged it down the river, away from the hive. As I swam, I was relieved to notice a decrease in the frequency of the bees hitting my hat; the buzzing was also becoming less loud and more distant. After a little longer I was finally free of bees, so I climbed into the boat, started the engine, and headed for Alejandro.

When I found him, he did not look like he was in very good shape. He was impossibly pale, although the skin of his face looked rougher and darker than usual: It was littered with bee stingers! I removed over 90 stingers from his face alone, and I could see a similar density of stingers scattered over the rest of his head. His hands and arms were also covered with stingers as a result of his trying to fight them off. I was certain that he had received no less than 300 stings. As we headed back his condition worsened: He was vomiting repeatedly, and his pale face gave the impression that he was about to faint.

Then I realized there was another problem I did not anticipate. To get to the high ground where we had left the vehicles, we had to cross a long stretch of flooded savanna that was not deep enough for the boat to go through. This part was too deep to drive across or even to wade through but not deep enough for the boat. There was a path in the savanna, a dry-season creek bed, which was the only path that was deep enough for the boat to go over. The bad news is that Alejandro was the one who knew the fine details of the path. His job as a tour guide at the ranch required him to use that path all the time, but I had only been in this section of the river a couple of times and could not begin to remember all the turns needed to stay in the deeper part and not run aground.

However, fearing that Alejandro was in real danger of anaphylactic shock due to the many stings, I had to do something. My sense of orientation told me the rough direction in which the cars were, and I headed there straight across the flooded savanna, ignoring the path. I will never know if the rains of the previous night had raised the water just enough for the boat to go over the flooded savanna, or if it was God himself who picked up the boat, but I was thankful regardless when I got to the trucks. Two hours later we were in the small rural hospital of Mantecal where the doctors, accustomed to this kind of emergency, treated him immediately and had him asking for lunch in less than 15 minutes!

Luckily, this story ends happily, perhaps more than you realize. Fortunately, Alejandro was not allergic to bees, and no further complications occurred. I wasn't sure if it was his hardy constitution or pure stubbornness that kept him from fainting, but much later he told me that he had prayed to God that if he survived this attack, he would give up his bachelor days, return to Spain, and marry his high school sweetheart. He kept his word!

Pregnancy and Parturition

Breaking Waters

Having gathered all the females, I was in a good position to get information on their reproductive biology. Parturition occurred at the end of the wet season from October to late December (Figure 6.11). Gestation lasted on average 202.6 days (standard deviation [SD] = 14.71). However, there was some uncertainty about the actual time when the animals were inseminated and when embryonic development begins. With a mating system that may last as long as 4 or 6 weeks, it is unlikely that the development of the embryos starts as soon as mating occurs, since it would mean a developmental difference of several weeks within the same clutch. It is likely that the sperm is stored during the mating and development does not start until mating finishes. Short-term sperm storage seems to be relatively common among snakes; some pit vipers regularly exhibit short-term sperm storage from fall to spring.[55]

No Pregnancy Cravings

Initially, I provided several species of fowl to feed the females that I had in enclosures while waiting for them to give birth, but they all refused to eat.

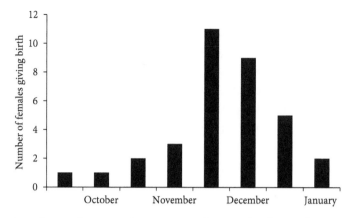

Figure 6.11 Seasonal timing of parturition of anacondas during the year.

Finding chickens to feed the snakes was one of those situations that helped me appreciate the effect anacondas have on people. I went to Mantecal, the closest town to my field site, where just about everybody has chickens in their backyard and many people are willing to sell some to any interested buyers. As I had been working on the ranch for a while, people knew what my business was: catching anacondas. When I asked the lady who ran the farm to sell me a couple of chickens, her reply was: "What for?"

This was an unusual question for someone who wants to do business. Since in the small town everybody knows everybody else, there was no point of my trying to dodge the answer. I told her I wanted them to feed a snake I had in captivity. The woman refused to sell me the chickens, arguing that she felt sorry for them. I tried reasoning with her that anyone else who buys chicken from her would likely kill and eat them, but that didn't seem to make a difference. She had identified with her chicken and she could accept the chicken being killed to feed people—she likely killed some regularly for her own consumption—but she could not accept the idea that her chicken would be used to feed a snake. I tried arguing that the snake was a pregnant female, hoping to appeal to maternal solidarity, but that didn't carry the day either. Being in need of chickens (or so I thought), I doubled my offer, but even then the woman refused to sell them to me. I ended up having a friend who worked on the ranch get the chickens for me, so I did manage to have food always on hand for the snakes I had in cages.

However, I was mistaken in my worries about feeding the snakes. During pregnancy, females did not eat anything. After they gave birth, though, they resumed feeding immediately, ingesting as many as eight chickens in a row!

We weighed a subsample of five animals three times during the pregnancy to monitor mass changes during gestation. The mass of the females barely declines through the gestation period. It is surprising that females did not lose more weight; in fact, one female even seems to have gained weight during pregnancy despite not consuming any food! This can be explained by consumption of water, to which the animals had free access during the gestation period. Clearly, fat tissue that the snake stored for pregnancy, as well as the yolk in the eggs, contains less water than muscle and connective tissue. As this stored fat was transformed into muscle and other tissues, adding water to them might have contributed to the relative lack of weight loss the anacondas experienced despite the very high metabolic expenditure associated with pregnancy.

It is possible that anacondas do not eat as a strategy evolved to prevent the risk of being injured in the process of killing a prey. An anaconda that attacks a prey while pregnant would expose her clutch to danger if she received a wound. In fact, long-term captive animals at Bronx Zoo, which are regularly fed with euthanized animals, ate throughout pregnancy until just 1 or 2 months before parturition (William Holmstrom, personal communication) and lost only 22% to 30% of their mass after giving birth.[56] This supports the idea that wild females are "playing it safe" when they stop feeding. Despite a very large breeding investment, female anacondas still have fat reserves after reproduction, as assessed in a few animals found road-killed after the delivery season and the relatively high CIA of some females after birth. Thus, the risk of a wound compromising the survival of the clutch is probably too high, considering that the anacondas may have enough reserves to survive. A similar behavior of little movement and no foraging during pregnancy has been reported in one other instance in a wild anaconda[57] and in other viviparous[22,58] and oviparous snakes.[59,60]

The fact that it is possible to predict pregnancy by using only using the condition index indicates that the main determinant for breeding in an adult female is whether she has enough fat reserves to do so. Some snakes rely on food they acquire during the pregnancy for their reproductive investment (budget breeders) while others use energy gathered prior to reproduction (capital breeder). The fact that anacondas do not feed during pregnancy as well as the fact that only those that have a certain body mass before breeding season reproduce in a given year suggests that anacondas belong to the latter group.

Maternity Room

The study of maternity required long hours of observation and "snake sitting." One of my associates, Rafael Ascanio, who was doing his undergraduate research, was key in this long and tedious follow-up as he did a lot of the legwork taking care of pregnant females and documenting what happened in one of the critical years. Gathering all our data, we collected information on the place where 19 births occurred either by witnessing them or by finding the remains of a birth event (stillborn, egg yolk, etc.). They occurred both on the land (12 times) and in the water (7 times). For 16 animals we managed to record the time of birth. Most births occurred in the evening after the hottest part of the day (Figure 6.11). Most births lasted between 20 and 40 minutes, but some animals took much longer (minimum = 10; maximum = 145). In three cases that lasted a long time, the females expelled some neonates or feces and then gave clear signs of distress, such as moving restlessly throughout the enclosure and spinning their bodies on the land or water. On two occasions when this happened there were a large number of stillborns, and we found that the females had some stillborns stuck in the duct. A normal birth probably takes between 20 and 30 minutes (Figure 6.12).

Learning things does not occur without some accidents. I tried my absolute best to take good care of the females in captivity. One reason was for the successful collection of data, as I needed to have healthy females for the

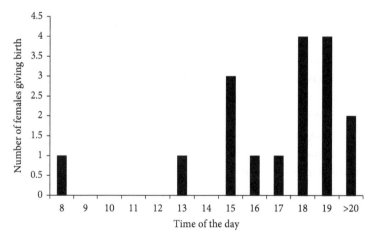

Figure 6.12 Time of the day when anaconda parturition occurred. All the births that occurred after 19:30 were scored as >20:00.[18]

study. But it was mostly because of the great admiration and love I have for the animals. It had been a long time since I had any snake in captivity, and I felt very lucky to have all these snakes for a good reason. It is possible, however, that some females were negatively affected by the handling and captivity. Some of them were very shy due to captivity and did not bask much, spending most of the time in the water. This might have prevented them from thermoregulating properly for the successful incubation of the eggs, so these females had a large number of stillborns and had poor breeding success. Not having enough exposure to the sun could lead to dystocia, the death of the clutch, or even the death of the female.[61] The skin of one female became very fragile, to the point of breaking when touched during handling. This animal died after having a completely unsuccessful clutch. This brought home to me how painful some learning experiences might be.

Females gave birth mostly in the evening or afternoon (see Figure 6.12). This might be driven by two reasons that are not mutually exclusive. One, a proximate explanation, is that the females need the high temperature of the day to trigger the energetic demands involved in the delivery, and another, ultimate, reason is that the neonates have a better chance of survival at nighttime when flying predators are not as abundant and the odds of dispersing safely are higher. Neonates tend to go into the water and very seldom use dry land.[17] The most likely scenario for wild anacondas is that the female gives birth in shallow water or at the water's edge, the same places where the female spends most of her time, and from where the neonates disperse. Babies do not linger around the female, nor does she show any parental care that I could observe. Both in the wild as well as in cages, neonates dispersed immediately after birth. However, I heard one account of neonatal anacondas in the Brazilian Amazon in which the clutch was seen near the female after birth (Laurie Vitts, personal communication). I also have seen neonatal anacondas in rivers in Cerrado (Brazil) basking nearby one another, suggesting some social behavior among neonates.

Stillborn Animals and Pregnancy

Normal clutches seem to have from none to two stillborns or infertile eggs. Most births (27 out of 34) had some stillborns and some had infertile eggs (11 clutches out of 34). Some eggs did not show any clear development, but during dissection some structures (resembling a small embryo) were identified, suggesting that some of the eggs were fertile but for some reason did not fully develop.

The mass of stillborns (average 214.68 grams [7.8 ounces]) was not significantly different than that of live ones (217.39 grams [7.7 ounces]). They looked fully formed physically and did not show any deformity or physical problem other than being dead. The reasons these individuals died before birth are not clear. It is likely that the female had problems expelling embryos and they asphyxiated in the womb. Animals that give birth late in the season seem to have a higher likelihood of having many stillborn offspring. This suggests that such females might have had some problems delivering or did not bask properly during pregnancy.

At least eight of the females observed during pregnancy ate or attempted to eat either stillborns, infertile eggs, or both. Right after birth the female pushes her snout across the mass of neonates, which encourages movement in those animals that did not crawl away right after birth. As she does this, she grabs and eats both stillborns and eggs. Several times a female grabbed a live neonate that was not moving, releasing it hastily when it moved (Figure 6.13). This makes complete sense. The female has fasted for the 7 months of pregnancy plus however long she did not eat before mating. Clearly, after 7 months of fasting and investing more than 30% of her body weight on reproduction she will be eager to recover some of the lost energy. Each neonate or stillborn represents a substantial amount of energy that she had invested in reproduction, and she can recover some of her investment by eating those that were not born alive. Plus, leaving the eggs and stillborns could attract predators that may follow the track of the live babies.

Wild-caught female anacondas had been reported to have problems giving birth in captivity in the literature. The researchers performed a cesarean section and found many stillborns in similar conditions to the ones I found; they also found a few individuals that were alive.[57] It has been reported in other snake species that individuals with clutches larger than average were at a higher risk of having stillborns or developmental anomalies.[62] In fact, pregnancy tends to be a critical moment for the health of the female. After parturition, some females often look very weak and thin and thus may be more likely to be the victims of predators (as in Olivia's case).[15,36,64] Animals in this condition may attack larger, more dangerous prey in order to overcome their energetic deficit, taking the risk of being injured or even killed by their prey (see Chapters 3 and 4).[24] The high cost of reproduction might be the reason that Olivia was so weak that she was killed by a caiman of the same size that she had eaten in the past. It was probably the reason that Penelope tried her luck on Ed and many other moments when a female takes a big risk attacking a very large prey. This is probably an important part of the reason that females suffer such higher mortality despite having a lot fewer predators than males (see Chapter 3).

In the Nick of Time

Anacondas give birth at the end of the wet season, a moment when the savanna has a lot of water and will not dry out in the next few months. However, as the dry season advances, the habitat for neonatal anacondas shrinks progressively. This is potentially not the best time for anacondas to have their babies. The peak of the dry season may be the least favorable for the survival of the neonates, since there is less water and the newborns have fewer refuges where they can hide from aerial predators. Furthermore, other predatory animals such as caimans, foxes, storks, herons, and tegus concentrate around the bodies of water. It seems that it would be more convenient for the anacondas to give birth at the beginning of the wet season so that the newborns would have a longer time to grow and reach a size that would enable them to fight off predators more easily.

However, being born during the time of peak flooding gives the neonates the most protection against predators at the time of their birth, when they are clueless and concentrated in the same place where their mother dropped them. Then they disperse quite a bit, occupying random locations in the landscape where the odds of finding them are rather low, as I have painfully learned. However, predation on the neonates is perhaps not the only selection pressure to consider.

The time when anacondas are born is the same time when there are a lot of young water-dwelling birds. If anacondas were born early in the rainy season, there would not be as many young aquatic birds that start mating with the rains and whose young start hatching later in the season. In the months of September and October, when most anacondas are born, there are plenty of aquatic birds for the babies to feed on. Furthermore, anacondas are not the only reptile of the llanos whose offspring are born at this time of the year. Spectacled caimans have a similar reproductive timing. In fact, it is possible that the timing of the newborn caimans might be a selection pressure contributing to the reported timing for anacondas. Newborn caimans use a habitat very similar to that of newborn anacondas and do not overlap in terms of the trophic niche since the baby caimans feed on arthropods and small fishes while anacondas feed mostly on vertebrates. A newborn caiman (weighing 40 to 50 grams [1.4 to 1.8 ounces]) represents between 20% and 25% of the mass of a newborn anaconda, which is just about the proper meal size for the babies. I have not found any newborn caiman among the diet of baby anacondas; however, the representation of this size class of snakes in my sampling is very small. Naive newborn anacondas seemed very interested in the cotton swabs that had been rubbed on newborn caimans in the predation trials done in captivity,[14] suggesting that they are part of their natural diet.[65]

The other side of the coin is the female's health and well-being. The timing of the birth may also be influenced by the best time for the female's survival. Since a female has a long life span and is expected to produce many clutches in her lifetime, after a large reproductive investment it is in her best interest to find food and recover from the breeding investment. Females that give birth at the end of the rainy season will have the next dry season when many potential preys will be forced to visit the bodies of water that the anacondas inhabit, where they can easily hunt them. It is also possible, though, that the timing of the parturition is a natural consequence of the timing of the mating, which may well respond to other variables (see Chapter 8).

Why Are Females so Large?

The more we learn about anacondas, the more interesting the question becomes of why they are so large. Clearly, since female anacondas are the large ones, they are a prime candidate to tell us the secrets of the evolution of their large size. Do female anacondas experience evolutionary benefits for being so large? The information I have gathered about their reproductive biology can shed some light onto this question.

All the Better to Have Large Babies

Neonates are fairly large at birth compared with the size of other snakes, including the other snakes known as "giant snakes" such as Old World pythons. Neonates of other large constrictors may weigh as much as 40 grams (1.4 ounces).[66] However, none of them compare with anacondas, which have an average birth weight of 217 grams (7.7 ounces; range: 150 to 330 g [5.3 to 11.6 ounces]). Contrary to what is true in the rest of anaconda biology, there is no sexual dimorphism among neonates, so the surprising sexual size dimorphism found in adults develops through the life of the individuals. Neonatal mass is strongly related to the SVL and CIA of the mother. This is not surprising, since larger females store more fat and can invest more per offspring, making them larger and increasing their chance for survival. The capacity of producing larger neonates would be a selection leading to large size in females since larger babies would have a better chance of survival than smaller ones.

All the Better to Have More Babies

Clutch size was calculated as the sum of all the live and stillborn neonates plus the infertile eggs. The average clutch size of the anacondas was 29.4, and it was strongly correlated with the size (mass) of the female. It is also correlated with the combination of SVL and CIA. In other words, the larger and fatter a female is, the more babies she will have, which is common among other large reptiles.[25,47,67-69] Not only are her babies larger, but she can make more of them. Larger animals have larger coelomic cavities and more room for the production and development of larger clutches.[70] It is clear, then, that females benefit from being larger so they can have more and larger babies, leaving more descendants for the next generation.

How Much do Anacondas Invest in Reproduction?

Anacondas invest a lot of their energy into reproduction, and the larger they are, the more babies they can have (Figure 6.14); but there seems to be a mixed strategy. The relative investment animals put into reproduction (the relative clutch mass [RCM]) can be calculated by dividing the mass of the clutch by the postpartum mass of the female (Figure 6.15). It is the proportion of the female's mass that is represented by the clutch. There was no clear relationship between RCM and SVL other than the fact that RCM was more variable in the smaller females than in the larger ones. However, removing statistically the effect of other variables, it becomes clear that the larger the females are, the smaller the total reproductive investment is. So, the larger a

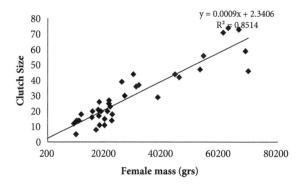

Figure 6.14 Relation between clutch size and the mother's SVL. The larger the females are, the more and larger babies they make.

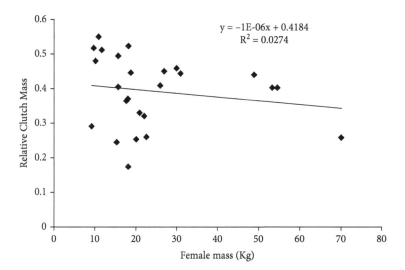

Figure 6.15 Relation between RCM and female mass. RCM is calculated by dividing the mass of the clutch by the mass of the females after birth.

female is, the more babies she has, the larger the babies are, and the less she invests in them proportional to her body size. All these are great benefits for a female to grow to a large size and may be a very important part of the selection pressure that makes females so large.

The RCM in large snakes was lower than but close to the average found for other species of snake[71] and within the value found for other large snakes.[25,72] The RCM is expected to be lower in viviparous aquatic species due to physical limitations for swimming. As the explanation goes, if the rear of the snake is too bulky and too heavy, as happens with a large clutch, animals would not be able to swim as well, so natural selection constrains the investment per clutch. It is not clear how a large body carrying many developing eggs would hinder an aquatic snake more than a terrestrial one, however. This has been well studied in seafaring snakes. If the caudal end of the snake is too bulky or heavy, it may hinder its ability to swim in open water.[71,72] However, anacondas, pregnant or not, do not move much and, unlike the sea snakes studied, anacondas often use shallow water with little current, so it is uncertain if a heavy lower body would really handicap their ability to move around. In fact, being aquatic relieves the snake from the need to carry the weight of her clutch and may make it easier for anacondas to carry larger clutches.

The average RCM in anacondas seems to decrease with the size of the snake (Figure 6.15). This is counterintuitive: Recall that young animals need to split their surplus energy into growth and survival and reproduction, so smaller animals are expected to devote less energy to reproduction than larger ones. This does not seem to be happening with anacondas. The fact that young animals investing more in reproduction does not support the idea that young individuals partition their stored energy into reproduction and growth. Such a strategy should have resulted in a small RCM in smaller sizes and large investment in larger sizes since larger animals do not forfeit future growth or breeding by having relatively larger reproductive output.[42] Instead, I found that the reproductive investment in young (smaller) animals is larger, and more variable, than that of larger animals.

A possible explanation for this higher reproductive output of young animals could be that smaller anacondas have a larger risk of being preyed upon,[63] so saving energy for future growth would not be adaptive and the best strategy would be to make very large breeding efforts when possible. This strategy apparently changes ontogenetically as the female becomes larger. There seems to be a threshold where the female switches strategies at around 30 kilograms (66 lbs), as suggested by the variance after that size.

The investment in anacondas is comparable to the investment found in other large species.[68] However, anacondas are at birth much larger than any of them; in fact, they are the size of many adult snakes. The female would get most benefit from making as many neonates as possible. Why is she not producing many more, much smaller babies? If the female was going to put, say, 30% of her mass into creating 30 250-gram (8.8-ounce) neonates, she would increase her reproductive output by, say, making 60 125-gram (4.4-ounce) babies. Other related aquatic pythons have babies weighing only 40 grams (1.4 ounces), so it seems as if anacondas could increase their reproductive output by having smaller babies. Why is she making babies this big even if it represents making fewer than she could theoretically make?

How Much Is a Baby Worth?

The neonates represent on average 1% of the mass of the female (Table 6.1), although there is considerable variation depending on the female's size. The reproductive investment per offspring is larger than 2% in smaller animals, dropping below 0.5% in larger females (Figure 6.16). It seems that animals smaller than 20 kilograms (44 lbs; smaller than 330 cm [10.8 feet]) tend to make substantially larger investment per offspring than larger ones. This reproductive investment per offspring drops as the animal grows larger; this is again contrary to what is expected. Smaller animals need to split their reproductive investment into growth, survival, and reproduction, so making larger babies leaves less energy available for growth and survival. Why are they making babies proportionately larger if this means making fewer of them? It is possible that baby anacondas need to be at least a given size in

Table 6.1 Body measurements of the anacondas caught and some statistics about reproductive output calculated from the parturition[15]

	Mean	SD	Min	Max	N
Total length (cm)	365.5	77.5	242.7	521.3	45
SVL (cm)	318.8	71.37	210.7	466	42
Mass (kg)	30.82	18.22	9.25	70.00	45
RCM	0.39	0.095	0.17	0.52	27
Relative investment per offspring (%)	1.03	0.43	0.32	2.15	32
Clutch size	29.4	18.4	5	74	39
Neonatal total length	78.77	4.21	63.9	91.43	504
Neonatal SVL	68.42	4.01	54.7	80.93	492
Neonatal mass	216.8	35.6	145	330	578
Sex ratio	1.25	0.59	0.38	2.6	21

See text for RCM formula. N = sample number. The sex ratio was calculated over the 21 clutches from which I have full data.

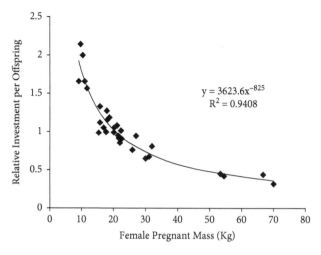

Figure 6.16 Relation between the relative investment per offspring and the SVL of female anacondas ($r = -0.899$; $p < 0.000$; $n = 30$).[15]

order to have a chance to survive (find meals, avoid predation, and so on). So small females need to invest whatever it takes to make babies that have a chance. Clearly it makes no sense to make a baby that is too small to survive given the ecological conditions it will find. This size seems to be around 150 grams (5.3 ounces), the size of the babies of the smallest clutch. This results in a relatively large investment per individual offspring.

It is quite possible that if the neonatal anacondas were smaller, they would have a very hard time coming across the food they need and they probably would have an even higher mortality. Diet data show that they prey on small birds, such as wattled jacana (*Jacana jacana*), and other animals ranging from 21 to 112.5 grams (0.7 to 4 ounces; average 76 g [2.7 ounces]). This represents 35% of the average neonate's size. If neonates were born at, say 125 grams, the average meal would be proportionally larger, perhaps too large for the animal's safety. Data on prey size from other taxa,[26] including neonatal anacondas,[13] show that they chose prey size around 25% of their own mass. Furthermore, they would have to endure that higher mortality for longer. Recall that they rely on prey that come within reach of their hideouts, so being larger allows them to consume a broader range of prey and, in fact, increases their odds of getting a meal. Of course, it does not matter how many babies a female has if none of them survive; her contribution for the next generations would be zero. Perhaps there is not enough suitable prey available to feed smaller neonatal anacondas. Or perhaps predation on a smaller anaconda is too intense and they have to endure it for so long that it would not pay off in terms of producing adults that bear their own offspring. So, natural selection had set the minimum anaconda size at 150 grams (5.3 ounces) at birth, even in relatively smaller females. If babies need to be a given size in order to survive, natural selection would lead females to make babies no smaller than that, even if it represents making fewer babies and even if that means a higher cost for the female both in total and per offspring.

Thus, the minimum size of the neonates might be a selection pressure influencing the size at which females breed. Having to make babies of a certain size might prevent females from reproducing sooner, since females smaller than a certain size will not be able to produce larger babies. This limitation might be an important part of the reason females are larger and perhaps take longer to reach reproductive size than their male counterparts. The fact that investment per offspring drops so fast with size among young individuals supports the notion that there is strong selection pressure to make neonates of at least a certain size. The need to produce large neonates is, thus, another reason selecting for large female size.

Newborn Babies

The sex ratio at birth calculated from all the newborns obtained in the study was nearly even, as seen earlier. However, the average sex ratio (female to male) from individual females is 1.25 ($n = 21$), which is significantly different from what would be expected by chance. So, while the total sex ratio of the newborns that are entering the population is close to 1:1, the average sex ratio of the individual clutches is not. Smaller females seem to have female-biased sex ratios, while larger females show a more even distribution.

In species that change reproductive output through life, natural selection acts to maximize the reproductive output. If the female can manipulate the sex ratio of her clutch, natural selection will select for the strategy that maximizes the number of grandchildren she will have. For instance, some sea bass are females when they are small and cannot outcompete other males. But some of them may turn into males after they reach a certain size or when the dominant male of the area dies or is removed. So, these fishes reproduce as females while they are small and switch to males if the conditions are right because males have a much larger reproductive output if they control an area with enough females in it. While this feat is uncommon and most species of vertebrates cannot change sex during their lifetime, natural selection can still act to maximize reproductive output using perhaps more subtle ways. A possible explanation for the larger production of females in small mothers is that males have a very small chance of breeding while they are small due to physical competition among males (see Chapter 8). So, because small females can only produce small offspring, the best way for the smaller females to secure some descendants in future generations (as in grandchildren) is by producing more females that will surely breed as soon as they reach adulthood.

Infertile Eggs

Surprisingly, the average size of the eggs (124.6 grams [4.4 ounces]) had no relationship to the female's mass ($r = 0.14$; $p = 0.69$; $n = 11$); instead, it was the growth rate of the eggs (mean 0.44 grams/day [0.02 ounces/day]) that was correlated with the female's mass ($r = 0.67$; $p = 0.023$; $n = 11$). Egg size is surprisingly similar among the females of varying sizes. There does not seem to be any clear tendency. If anything, there appears be the same pattern that we see in the RCM, where smaller females have a larger variance than larger ones. While the neonates from larger females are larger than those from smaller ones, this trend is not present in the eggs. This involves a differential growth rate in embryos of females of the different sizes. The difference in size of neonates that developed from eggs of equivalent size suggests a differential rate of turning over stored nutrients into tissue, perhaps involving the participation of some sort of placenta.[73]

Why do larger females produce relatively smaller eggs than small females? Allometric growth of the reproductive organs would predict that the embryos from large females would be proportionally larger.[74] Furthermore, production of a placenta for the development of eggs, plus additional provisioning of it, means an extra investment of energy and structures. It is possible that the females do not invest in larger eggs so as not to expend too many resources on some eggs that might not be fertilized, since the females do not seem to be able to reabsorb the energy invested in the unfertilized eggs. The fact that some eggs were fertilized but did not develop could involve two different phenomena. First, there could be incompatibility between the sperm and the egg;[75] second, the females might exert control over the paternity of her clutch by selectively developing some eggs (see Chapter 8). Both of these are fascinating possibilities that deserve further attention.

Pros and Cons of a Large Size

In summary, we find that female anacondas obtain a lot of benefits from being large. Large size is normally constrained by the high cost of mobility associated with large mass and a conspicuousness that prevents the animal from stalking prey and makes it vulnerable to predation. Anacondas overcome the first two thanks to their aquatic habits. The aquatic environment allows them to glide effortlessly despite their large mass and also allows them to stalk prey hidden under the aquatic vegetation. Thanks to these two conditions anacondas can overcome the third, the problem of being seen by predators. Very large anacondas simply outgrow predators, making them free to obtain all the other advantages that large size brings to an individual. Another problem that large animals often have is obtaining enough energy to support their large body. The rich fauna found in the area provides abundant food to sustain a large biomass of predators, and due to their ectothermic metabolism and low activity levels, anacondas may be able to get by with surprisingly little energy intake compared with equivalent-sized endotherms.

As for the advantage of large size, all seems to be on the side of female anacondas. Large females produce more offspring, of larger size and higher survival, investing less energy per clutch and per individual offspring. Larger babies suffer less predation, have access to more and larger prey, and reach reproductive size sooner. Furthermore, larger animals (mothers and babies) are better able to survive times of shortage because they may have more reserves to survive fasting. All these are selection pressures for large size in females. These benefits for large size beg the next question.

Why Aren't Anacondas Larger?

In fact, given all these benefits for large size, one cannot help but wonder: Why aren't anacondas larger? Notice that fecundity of females does not plateau at the larger sizes (Figure 6.14). This suggests that females might benefit by reaching even larger sizes. Why don't they? Is there any limit on how large

an anaconda can grow? Are there ecological or physiological limits that con-
strain female size?

Spying on the Breeding Females with the CIA

I have already described the CIA and how it helped me know if a female was
going to breed or not. Contrary to the trend found in the general population of anacondas, where the average CIA remains the same regardless of the
animal's size, the CIA of pregnant females decreases with size (Figure 6.17).
This indicates that larger animals that are breeding are relatively thinner than
smaller animals. This is important because this decrease of CIA among pregnant females is not a consequence of allometric growth but is a true ecological trend in female reproduction. In other words, the lower CIA in larger
pregnant females is not because anacondas become thinner as they grow; it is
a pattern that is limited to pregnant females. As females become larger, they
seem to breed with relatively less stored fat. The beauty of this is that it seems
that if a female grew to a certain size, she might be so large that she would not
reach the condition needed to reproduce. It seems like a female may reach a

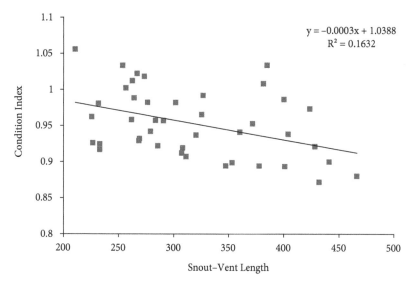

Figure 6.17 Size-related changes in the condition index before birth of
pregnant female anacondas in relation to female SVL ($r = -0.42$; $p = 0.007$;
$n = 41$).[15]

size where she will not meet the threshold (0.875) needed for reproduction (see Figure 6.8).

Why do Females Get Skinnier as They Grow Larger?

Is it possible that the larger females may be constrained by their mass? A very heavy snake might be constrained when crawling on dry land, which anacondas must often do in the seasonal savanna. Thus, as the female grows in length, there is a maximum size she can reach and still be able to carry her body on land. Not having a ribcage (for lack of a sternum), they need to lift their ribs with every breath. There may be a mass limit where the weight of the ribs is too large to lift in every inhalation while the animal is on land. This consideration might set an upper mass limit that forces the female to grow thinner as she grows longer. This is the reason that stranded whales suffocate on the beach. Even though they can breathe air, their weight on land collapses their lungs. Thus, anacondas that live in hyperseasonal habitats, such as the llanos, may face a similar problem when trying to crawl over land. If the female is too large (or too fat), she might not be able to spend a lot of time out of the water or may be limited to staying in the water.

Alternatively, a lowered CIA of larger animals may be a consequence of the fact that larger animals only need to gather enough energy to breed, and they need no reserves for future growth. Thus, they can breed as soon as they meet the minimum surplus of energy to do so. However, they would benefit from breeding with a higher CIA, since fatter females make larger clutches. Also, ending reproduction with more energy stored would allow them to better recover from the long months of fasting and rebreed sooner. The drawback of this is that the anaconda would have to postpone reproduction for another year (or more), and during this time she will still face risks such as drought, disease, and the ever-present risk of injury while taking a prey. Because of the unpredictability element, the anaconda benefits from breeding as soon as she is ready to produce a clutch, relatively thin as she may be, instead of taking the extra chance of not making it to the next breeding season.

Another explanation from a different level of analysis (not mutually exclusive) is that larger females might take longer to gather the energy they need and, thus, they need a longer time to attain the required CIA. This happens in water pythons (*Liasis fuscus*). They have problems gathering the amount of energy they need to be able to breed, and so the larger individuals do not

breed as often as the younger ones, and the largest might not breed at all.[68]
Due to the large amount of potential prey found in the llanos (capybara,
caiman, turtles, and wading birds) in the very areas where the anacondas
occur, it is hard to believe that food availability might be a limitation.
However, consumption and processing of large prey items may take a long
time, extending the time females need to obtain energy.

Larger anacondas have a different diet than smaller ones. The latter feed
primarily on birds and the larger ones feed mostly on mammals and reptiles.
Anacondas have the added cost of healing from the wounds that larger snakes
are more likely to experience (see Chapter 5). To feed on larger prey might
involve a lower rate of feeding, since it takes longer to catch, process, and
digest them and larger prey are less common than smaller ones. Although
larger snakes have lower relative energy expenditure per unit of body mass,
their absolute metabolic expenditures are considerably larger than those for
smaller ones.[31]

Because larger females take longer to accumulate mass, it is possible that
as soon as they reach the threshold for reproduction, there is a mating season
coming along and they breed as soon as they can. This is convenient because
they secure that bunch of babies and do not risk mortality for another year.
Clearly, if the female postponed reproduction and died before the next year,
she would forfeit the 70 or 80 she could have borne had she bred as soon
as she was ready. On the other hand, younger females, which can accumulate energy more quickly, may pass the threshold needed for breeding in the
season. For example, if the female reaches the threshold in July, when there
are a lot of birds and young capybaras, she will continue accumulating energy until the following March or April, so these females would be well over
the cutoff for reproduction, while the ones that take longer will always be
very close to the threshold.

Courtship and mating itself can play an important role in inducing
ovogenesis.[76] This phenomenon might happen more with larger females
than with smaller ones, since larger females are more sought out by the
males (see Chapter 8). Support for this idea is that Olivia (a very large female) mated in 1995 with a lower condition than the threshold for reproduction (8.5, short of 8.75). I found Olivia the following year extremely
weak, with many capybara wounds. Shortly afterwards, she was attacked
and killed by a caiman no larger than another caiman I had seen her eating
(see Chapter 1[63]). If the ovulation is induced by the male's courtship or
mating, then it is possible that a female who is marginal enough to breed

can be induced into an inconvenient breeding event. Olivia was a very large female, with her skin all covered by scars. She did not grow at all in the 5 years that I followed her and was, perhaps, very old. She gave birth in the wild so I could not collect reproductive information such as RCM or clutch size. The question remains as to whether she made an extremely large breeding investment (perhaps suicidal since her future reproductive potential was low at her old age) that produced her weak state, or whether she was simply too thin to breed (as suggested by the CIA) and mating in that year was a wrong decision. The idea of a maladaptive mating induced by courting males contradicts the conventional wisdom that female snakes emit pheromones to attract the males.[77,78] However, it is a working hypothesis for a species that may not be fully adapted to the llanos and might have evolved in different environments (see later in the chapter).

How Good versus How Many

The decrease in CIA suggests that larger females may not accumulate fat as easily or as quickly as younger individuals. If this were the case, we should expect that the breeding frequency would drop as the female grows larger. This would not be convenient, however. Clearly, since natural selection favors those individuals that leave more descendants, there must be strong selection to increase breeding frequency. I set out to see from my data if there was such a decrease in the breeding frequency in larger individuals. However, to know when an animal breeds is not straightforward; it occurs in the field and animals are not always easy to find.

I used two methods to assess breeding frequency. One was by determining the condition index of all the animals caught during each year and then calculating the proportion of animals that bred in that year. This cross-sectional method offers a large sample size but is sensitive to detection bias: Breeding females were more conspicuous than non-breeding ones, and thus it is likely we were overestimating breeding frequency. The other estimate was based on those animals that I had caught in consecutive years and thus would know if they bred or not. This way I could estimate a breeding frequency for the population by averaging the breeding frequency I obtained from individual animals.

Both methods show that the breeding frequency decreases as anacondas become larger. The proportion of breeding females in the cross-sectional

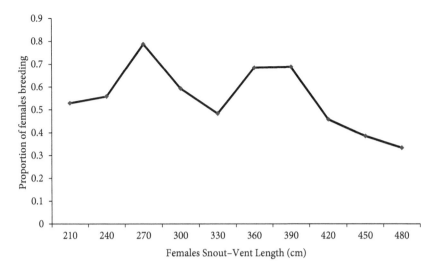

Figure 6.18 Proportion of pregnant female anacondas in different size classes. Calculated cross sectionally from the capture of the total sample ($n = 222$) by using the condition index.[15]

study was 0.57, but it varies across different sizes. Breeding frequency is inversely correlated with size (Figure 6.18). On the other hand, the longitudinal follow-up of marked individuals shows an average of 0.379 (Figure 6.19). This seems to be a combination of some animals in a biannual cycle (0.5) and some that breed at a lower frequency. Larger animals seem to take a longer time to recover and engage in a new reproductive event. Some animals skipped reproduction for 3 years and bred in the fourth (as in the case of Olivia), and some other animals did not attain breeding condition for 4 straight years (as in the case of Chinga). Chinga was missing the tip of her snout, which was an open wound that did not seem to heal. This probably affected her hunting efficiency, and she might not have been able to gather the energy needed to breed, although she did survive all those years, which suggests that she was getting enough food to get by.

As expected, the cross-sectional study seems to overestimate breeding frequency. Although the results of these two methods are not fully in agreement as an average breeding frequency for the population, they both show a negative correlation between breeding frequency and female length, consistent with my hypothesis that larger females may take longer to accumulate the energy needed for reproduction.

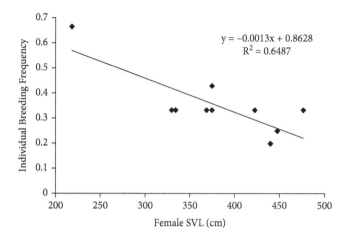

$$y = -0.0013x + 0.8628$$
$$R^2 = 0.6487$$

Figure 6.19 Frequency of reproduction of female anacondas that were caught in several consecutive years and were breeding in at least one ($r = -0.805$; $p = 0.005$; $n = 10$).[15]

One aspect that seems clear is that anacondas do not breed every year. This is not surprising since females give birth at the end of the year (see Figure 6.11). Since their mating occurs in the dry season (March and April; see Chapter 8), a female anaconda would have only a few months to recover from the reproductive effort. Considering all the effort required to find and catch a prey, it is not surprising that the few months between parturition and the following breeding season are not enough. The most frequently anacondas seem to be able to breed is every other year. The only two cases in which the data suggest that anacondas might have bred in consecutive years, based on the CIA, were in very young individuals that perhaps skipped reproduction in the first year. The criterion I used to determine adulthood was the size of the smallest female ever found breeding. It is possible that some of these females were already longer than this smallest size but did not happen to reproduce that year. They might still have taken another year before engaging in their first reproductive event (Table 6.2). Private breeders and zookeepers have noted their failure to breed animals that eat *ad libitum* in captivity in 2 successive years (William Holmstrom and Peter Strimple, personal communication). A biennial pattern in reproduction has been suggested or found in several snake species.[22,58,79] A decrease in breeding frequency with size also has been reported,[36,68,80] and this even seems to be the rule in large-sized species.

Table 6.2 Breeding frequency of the adult female anacondas that were caught in several years[15]

Year	1992	1993	1994	1995	1996	1997	1998
E101C	Y	—	Y	—	—		
Lina	N	Y	—	—	—	—	
Kathalina	N	Y	—	—	—	—	
E145C	Y	N	—	—	—	—	
E155C	Y	—	Y				
E161C	N	Y	—	—	—	—	
Hermelinda	Y	N	—	—	—	—	
Sarah	Y	—	—	N			
Laura	—	Y	—	Y	—	—	
Araine	—	Y	N	—	—	—	
E200C	—	N	—	N	—	—	
E223C	—	N	Y	—	—	—	
Renée	—	—	Y	—	N	—	
Diega	—	Y	—	—	—	Y	
E436C	—	N	N	—	—	—	
E486C	—	—	Y	—	Y	—	
Guaratarita	—	—	Y	—	N	—	
E78C	N	Y	—	—	—	—	
Julia	—	—	—	—	Y	Y	
Mary-Jo	—	—	—	—	N	Y	
Alice	—	—	—	N	Y	—	
Courtney	—	—	—	N	Y	—	
E90C	Y	—	N	—	—	—	
Yuang-Ly	—	—	—	N	N	Y	
E204C	—	N	Y	Y	—	—	
Zuca	N	N	—	—	Y	—	
Musiua	N	N	—	—	Y	—	
Mirna	Y	N	N	—	—	—	
Mónica	Y	N	—	—	Y	—	
Antonieta	—	Y	—	Y	N	—	
Andrea	—	Y	—	Y	N	N	
Marion	N	N	Y	—	Y		
E437C	—	Y	—	N	N	Y	
Chinga	N	N	N	N	—	—	
Joan	—	—	N	Y	N	N	
Olivia	N	N	N	Y	N	—	
Madonna	Y	N	—	N	N	Y	N
Sue	Y	N	N	Y	N	Y	

Y = yes, N = no.

Can a Female Be Too Large for Her Own Good?

Females seem to experience benefits from being large in terms of larger clutches and larger offspring, but larger individuals also have a lower breeding frequency. There seem to be a tradeoff between clutch size and breeding frequency. The breeding frequency decreases with the size of the female, so a point might be reached where the benefit of larger clutches might be balanced out by the decrease in frequency. The optimal situation would be for the females to have a large size so they can have larger clutches but to remain small enough to find food, locomote easily, and breed regularly. Can females grow too large for their own good?

I calculated the yearly reproductive value of a female by multiplying the clutch size times the calculated breeding frequency of a female due to her size. Thus, it represents the number of neonates the female produces per year, if she breeds every year. What I learned is that the yearly reproductive value of females increases with body size despite the decrease in breeding frequency (Figure 6.20). So, based on the animals for which I have data, there doesn't seem to be any problem in the female growing so large that her size lowers her yearly reproductive output.

So, why aren't the females even larger? Figure 6.20 shows that females benefit from ever-larger sizes. Clearly, since I have not caught any animal larger

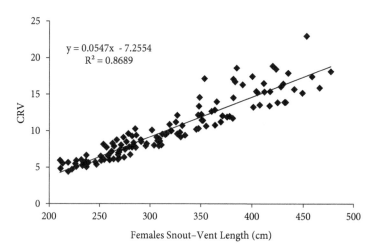

Figure 6.20 Relationship between the yearly reproductive value (breeding frequency multiplied by the clutch size) and SVL of female anacondas ($r = 0.902$; $p < 0.000$; $n = 129$). CRV = current reproductive value.[15]

than 5 meters (16.4 feet) SVL, I do not have empirical information about the yearly reproductive value of females longer than 5 meters SVL. As I said earlier, I strongly prefer results that come from the field, the ones that I sweat for and endure heat and thorns to obtain, over results one gets from a computer. However, not having data from anacondas longer than 5 meters, I had to resort to mathematical models. To address this deficiency, I modeled the yearly reproductive value of anacondas of larger sizes. Using regressions from my data, I extrapolated what would be the yearly reproductive value of an anaconda larger than those I have data for (Figure 6.21).

Here I had another opportunity to be humbled by the benefits of integrating field data collection and mathematical modeling. Figure 6.21 shows that the yearly reproductive value of a female peaks at about 480 centimeters (15.7 feet) SVL and decreases after this size. If a female were to grow larger than that peak, her yearly reproductive value would decrease. It so happened that the largest anaconda I have data for was 477 centimeters SVL (see Chapter 4). So, it seems that if females grew larger than 5 meters SVL, the number of babies they produce every year will decrease as a consequence of taking too long between breeding events—so they don't! Figure 6.22 summarizes the selection pressures affecting female size.

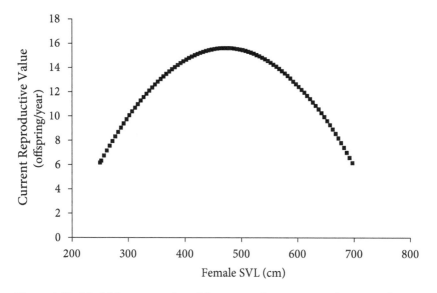

Figure 6.21 Model for anacondas of the expected ontogenetic change in the current reproductive value (CRV). The CRV is calculated by multiplying the expected breeding frequency by the expected clutch size.[15]

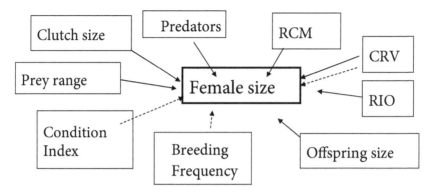

Figure 6.22 Selection pressures affecting female size in green anacondas. Those selection pressures that increase female size are represented with a solid line. Those that limit or decrease female size are represented with a dashed line. RIO = relative investment per offspring, RCM = relative clutch mass, CRV = current reproductive value. CRV represents the yearly clutch size of the animal in a given size and selects for an optimal female size; thus, it is represented with double arrows.[15]

Fair-Weather Breeders

Since breeding is tightly linked with fat storage, I surmised that there may be a relationship between breeding frequency and weather conditions. Clearly, in those years where there is more rain, there would be more herbivores (deer and capybara), which are staple foods for big anacondas. Also, more rains will favor insects, crustaceans, and fishes, which represent the base of the trophic pyramid that sustains birds and other anaconda prey. Thus, it is reasonable to expect that the anacondas would be better able to reach the breeding condition when their prey levels are higher.

However, as I started exploring this with my data and the precipitation records, I was surprised to see that breeding frequency is poorly correlated with precipitation (Figure 6.23). It occurred to me that, just like the anacondas need at least 2 years between breeding events, it might make sense to look at the effect of the precipitation from the last 2 years on a given year's pregnancy rate. Bingo! Figure 6.24 shows the pattern where females' pregnancy rate goes up with more precipitation.

What are the mechanisms by which precipitation affects pregnancy? Precipitation certainly has an effect on plant growth. Plants are the basis of the trophic chain that feeds most animals one way or the other. Grass feeds

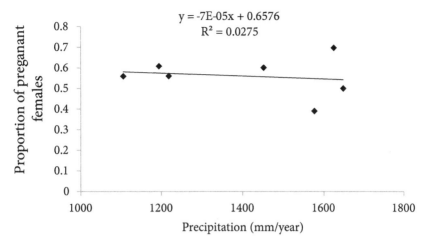

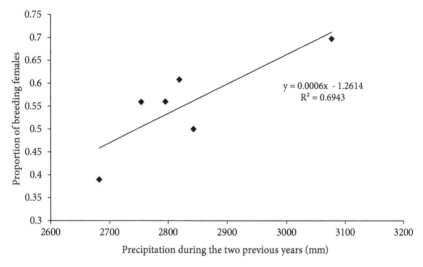

Figure 6.23 Proportion of breeding female anacondas each year in relation to the total precipitation for that year.

Figure 6.24 Proportion of breeding female anacondas found in relation to the total precipitation for the two previous years ($r = 0.66$; $p = 0.16$; $n = 6$).

herbivores and herbivores feed carnivores. An increase in grass increases the biomass of different levels of the food chain, eventually reaching the anacondas. In our case there are two main players that feed on grass. Capybara and white-tailed deer appear regularly in the diet of anacondas, and they are good candidates to explain the increase in anaconda pregnancy caused by increased precipitation, provided that we allow a year or two for the energy to flow through the trophic levels.

As it turns out, it is possible to measure the amount of grass produced in a given place by using satellites that regularly take pictures of the planet. By looking at the radiation of a specific wavelength in which grasses reflect the sunlight, we can determine the amount of grass in the savanna at the peak of the dry season and then at the peak of the wet season. The difference can tell us how much new grass was produced during that year.

Capybaras, cute as they are, have traditionally been the object of commercial management. As the story goes, one year a priest was asked if it was OK to eat capybaras during Lent (strict Catholics do not eat meat during Lent). Because they look much different than a cow and are always in the water, this priest felt they were *fish enough* that it was OK to eat them during Lent! So, the consumption of capybara during the Easter season has been traditional in Venezuela for more than a century. In the mid-1980s, the Venezuelan government started to monitor populations of capybara for its sustainable management, so during the entire time of my study there have been data on capybara abundance. Having all the data, I simply had to plug them into a mathematical model and see the results.

I found what I was looking for, but by chance I also found some information I did not expect. For starters, the model predicts more than 98% of the variance found in anaconda breeding frequency. In other words, the variables that I entered into the model can reasonably be considered as determining anaconda pregnancy (Figure 6.25). But it is also possible to disentangle the

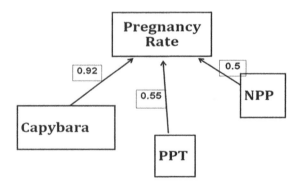

Figure 6.25 The forces determining reproductive output (yearly pregnancy rate) in green anacondas. PPT = annual precipitation, NPP = net primary productivity.

contribution of each variable on another by using statistical methods. I was able to confirm that abundance of capybaras alone was enough to explain a significant amount of the variation in pregnancy rates. This is not surprising given the number of capybaras in the anaconda diet. But it isn't only capybaras that play a role. Figure 6.25 shows all the variables that have an independent effect on anaconda pregnancy rates. We see that capybara abundance has a strong role in determining anaconda pregnancy, as indicated by the coefficients. But the figure also shows how abundance of grass and precipitation *alone* have a strong influence on anaconda pregnancy that is not via capybara abundance. Because the model has removed the effect of the other variables, the influence we see of grass on anaconda reproduction is not via feeding capybaras. That is accounted for in the model. In addition, the influence of precipitation on anaconda pregnancy rate is not via producing grass for herbivores but it is an independent effect instead.

A beautiful thing about science is that one answer always leads to more questions. How are precipitation and grass independently affecting anaconda pregnancy if it's not via feeding capybaras? I don't have any measures of the abundance of white-tailed deer in the area, so I can't enter their abundance as a variable in the computer model. However, since they are the single other herbivore that is important in the anaconda's diet, it is safe to assume that deer abundance is also a major contributor for anaconda pregnancy. A similar conundrum exists in the influence of precipitation that is independent from production of grasses. Again, I can only guess from the data, but a possible explanation is that precipitation increases the habitat available for aquatic organisms, increasing net primary productivity from plankton, which feeds a completely different trophic chain than the one that feeds on grasses. It goes to invertebrates, aquatic birds, caimans, and turtles, all of which form part of the dietary niche of the anaconda. Also, the increase in habitat may produce non-trophic effects that may also favor anacondas. These are new ideas to test and hypotheses to work on. There is not a single boring day in the life of a scientist!

Another Friend

The influence of capybaras on the anaconda's reproduction and life history is striking. So intimate is their interaction that we need to consider them in just about any discussion that involves anacondas. On a much less serious note, I will refer to another friend that I made when I was living in the llanos: Hairball. She was one of three newborn capybara that was found alone in the field after a male had killed some of her siblings from a rival male. Infanticide is a common phenomenon among all mammals, and capybaras are no exception.

Viviana Salas, a colleague who was sharing the field station with me while she studied capybara behavior, rescued them, feeling sorry for them since their mother was gone (not unlike the way I prevented Olivia from being killed by the caiman) (Figure 6.26). Two of the neonates were so badly wounded in the attack that they did not survive despite our care. When Viviana left after completing her fieldwork a week or so later, I was left with the duty of caring for my new friend. Small as she was, I had to feed her with a bottle, and often milk spilled over her throat and chest (Figure 6.27). Chuka became very adept at licking her up and down, cleaning up any remnants of milk left on her coat. This treatment seemed to evoke a piloerection response on the young capybara: Her hair stood up and she became very much a hairball. Being the only quadruped of the right size in the house, Hairball kept trying to nurse from Chuka, which she, of course, did not find very amusing at all (Figure 6.28).

Hairball grew up to be a full-sized capybara while I did my work. Unlike Chuka, she didn't follow me into the field but spent the day grazing in the area around my house until I came back (capybaras have very well-defined home ranges). Since she was so tame, the tourists loved her because they could take pictures of themselves petting a real-life capybara. So, when my time to leave the ranch came, I wasn't nearly as worried about leaving her as I was about Chuka. She was very welcome on the ranch: There were lots of wild capybaras around, so she could join their groups or stay a pet as she pleased. My friends who worked there told me that she hung out around my house for a while after I left and eventually joined a local group of capybaras that lived around the houses. They could always tell her apart because of her ear tags.

I was happy for her. I returned in later years but could not find her in the brief time my visits allowed me. Because 5 years is about as long as capybaras live, I didn't have any hope of finding her. I was happy to know she had survived and had had a happy capybara life, surrounded by others in a place where capybaras are really spoiled—especially if they are tame and live in a place where they aren't prey (Figure 6.29).

Some 5 years later I returned for a visit with a film crew from *National Geographic*. As I walked with my friend Ed George ("Gringo Loco") to find a spot for an interview, we approached a tree that provided just the right shade for the interview with a nice backdrop that Ed wanted. The place was so cool that there was a family of capybaras there passing the heat of the day. As we approached, the capybaras got up and left. This is common behavior. They don't flee because they are used to hanging around tourists, but people walking directly up to them they don't like.

However, as they were leaving, I said something to Ed. Because I was following the capybaras with my eyes, I noticed that one of them stopped in her tracks the minute she heard my voice. I felt a rush in my stomach thinking it could be Hairball after all these years. I called her, making a familiar whistle that she had learned when she was little. The 40-kilogram (88-lb) capybara cantered toward me, having recognized my whistle! She no longer had the ear tags she wore as a baby, but it was all her. I petted her and after a couple of strokes on her back, she lay down next to me to have her tummy rubbed, just like in the old days (Figure 6.30).

Two Strategies

Up to this point, I have assumed that size and age are perfectly related, but this is not necessarily the case. Old individuals that have had a low energy income can stay relatively small, while relatively young animals that eat well can grow to larger sizes in a short time. Thus, differences in the behavior of animals associated with their individual metabolic activities can produce confusion in these trends.[47,81,82] Differences in how the animals invest their energy can produce even more confusion. There might be some animals that invest a lot of energy in reproduction and might not have a lot of energy left to grow.

These differences could be genetically determined, developed ontogenetically as the animal grows, or both. Thus, there might well be older individuals that have not grown much due to the fact that they invest all their energy in reproduction and very little in growth. These animals would be displaying the optimal strategy for habitats with low resources and high predation. There may be animals that breed less often and acquire a larger size because they devote a larger part of their resources to growth. These animals would be displaying an optimal strategy for habitats with lots of resources and low predation pressure. These differences in life history strategies would account for the large variance found in smaller sizes, since smaller individuals could be young animals with a "grow large" (GL) strategy or older animals with a "high reproduction" (HR) strategy. The difficulty in assessing the ages of reptiles without using invasive techniques prevents an easy solution to this question.

The presence of some relatively small animals that did not grow over several years suggests that individual differences in growth do exist. Growth rate varies tremendously from individual to individual (see Chapter 4). The presence of two different tactics coexisting under similar conditions has been documented. It has been found that that some captive-reared *Elaphe guttata* that were fed a high-energy diet bred at 20 months of age while other individuals raised under the same conditions did not breed until 32 months of age. The earlier breeders had a lower lifetime reproductive success than the later breeders.[81]

Of course, the question of age can be also a relative one. For humans it is pretty simple: You count the number of days and years since your birth and that is your age. In human standards, time makes direct sense and it is directly correlated with a number of very important physiological variables (growth, teething, sexual maturation, etc.). A child of, say, 4 years has to be within a certain range of sizes; growth cannot be slowed without irreparable harm to the child's development, and it cannot be expedited a whole lot either. This might not be quite the case with anacondas. An anaconda can retain a childlike size for years if it had not been well fed, or it can reach an adult size in a very few years, again depending on how well it had been fed. Furthermore, the strongest aspect in anaconda biology (sexual size dimorphism, with very small males compared to females) can even be masked if the males are overfed (see Chapter 8).

Maybe to talk about an anaconda's age we need to find a snake-centric approach. It is quite possible that if you asked an anaconda her age she would answer: "I am 12 jacanas, 37 ducks, 5 turtles, and 4 capybaras old." The point is that time does not seem to be as important in an anaconda's biology as it is for us, or as food consumption is. Animals practice different energetic strategies, and this might lead to different growth rates that will have repercussions on how much the animals grow, and how much and when they reproduce.

Evidence of Two Strategies

Figure 6.18 shows how the breeding frequency changes with the size of the female. It has a potential weakness, though, because the breeding females may be easier to find. If we assume that this bias is similar across sizes, we find an interesting piece of information from this figure. Notice that there is a pattern that does not match the shape of the size distribution of the total population (see Chapter 4). There seem to be a dip between 300 and 350 centimeters (9.8 to 11.5 feet) SVL, as if females of this size did not breed as often as smaller or larger females. This seems to be an important time in the life of an adult female anaconda.

A possible scenario is that females reproduce as soon as they can make viable babies to secure some offspring. Youngsters still may suffer predation from some mesopredators of the swamp. So, at this point it is best to secure some descendants. Once the female surpasses 350 centimeters (11.5 feet) SVL, she is pretty much predator free. At this point, it may be a good investment for her to skip reproduction and use that energy for growth since her odds of being preyed upon are smaller at a larger size. Further support for the idea of a switch in strategy exists because anacondas switch diet from birds to mammals and reptiles in the same size bracket. When individuals switch their diet, females might go through a period when they cannot find the right amount of food due to their lack of experience in finding, subduing, or ingesting it. All the energy consumed may be put into growth in order to increase her fecundity, so the female may skip reproduction for a few years.

Another possibility is that some animals breed and stay small throughout their life, expending all their energy on reproduction and putting none into growth after reaching minimum reproductive size, while others grow to a larger size and start reproducing past the 350-centimeter bracket. These two strategies may be successful in their own way, and thus each stays in the population as a stable evolutionary strategy. In fact, should there be a place where green anacondas get anywhere close to the 10-meter mark, their biology will be that of GL individuals that start breeding at a larger size and their natural history would be different than the HR strategy that may be favored in the Llanos.

Why Would Two Strategies Exist?

Two alternative strategies may coexist if both are equally efficient. The contribution to the population of individuals smaller than 340 centimeters SVL (first peak) and of individuals larger than 340 centimeters SVL (second peak) may be calculated by adding up the estimated clutch size of all the breeding females of those two size groups. The second peak, despite the lower number of females, is responsible for 59.5% of the offspring born in a given year. Furthermore, the offspring of larger females have a higher expectancy of survival due to their larger size, so their contribution to the recruitment in the population must be even higher. If GL females produce more offspring than HR ones, they are expected to replace them in the population. However, the smaller animals may secure enough offspring in the population that the alleles leading to this strategy (if inherited) are not eliminated. Also, climatic fluctuation may mean that one strategy is favored at sometimes and the other at other times.

HR animals (and a slow growth rate) might produce higher contributions to the population than GL individuals during times of prolonged shortage (several strong dry seasons or relatively dry wet seasons) when the larger animals have problems finding food and are more constrained by the drought. GL individuals might do better in wetter periods. Changes in the efficiency of these strategies between the years, due to environmental differences, might prevent either of them from replacing the other in the population.

Alternatively, the genetic flow from other areas might prevent adaptation to local conditions. The savanna (*bajío*) is adjacent to the riverbanks, and the seasonal overflowing of rivers is part of the water supply of the savanna. There may be gene flow between these populations; however, the habitats are quite different, and the selection pressures are expected to be different too. Animals with a GL strategy are probably adapted to the river and animals with an HR strategy may be fitter for the savanna. The shape of Figure 6.18, the distribution of pregnant females found, may result from two strategies adapted to two different habitats. Anacondas in the savanna live right next to the gallery forest and the river. The sample (collected in the savanna) is probably composed of some HR individuals (first peak) and fewer GL individuals as well, which are fittest in the neighboring river.

The constant migration of animals (or genes) from one side to the other prevents the population from fully adapting to the conditions they are living in. This lack of adaptation to the local conditions due to genetic flow from a neighboring population has been reported to occur in other species.[83,84]

7

How Big Can a Giant Be?

Of the BOAS.

It was well known among all the Romans, that when Regulus was Governor or General in the Punick wars, there was a Serpent (near the River Bagrade) killed with slings and stones, even as a Town or little City is overcome, which Serpent was an hundred and twenty foot in length; whose skin and cheek bones were reserved in a Temple at Rome, until the Numantine war.

From *A History of Serpents* [85, p. 671]

The maximum size that anacondas can reach has been the subject of long-standing debate among herpetologists. There are many accounts of snakes around 9 to 11 meters (29.5 to 36 feet).[86] A lot of the controversy concerns the credibility of the records, the confusion created by the fact that the skins stretch when the snakes are skinned, and animals whose length was merely estimated or measured with unreliable methods. I do not intend to revisit that discussion, or review all the reports, but it is striking that the largest snake that I have caught, out of nearly 1,000 animals, is only a little more than 5 meters (16.4 feet) long. What is the reason for such a difference? Why have I not found any animal anywhere near 9 to 11 meters—or close to 6 meters (20 feet), for that matter?

Due to their slow growth rate (see Chapter 4), anacondas require a long time to reach a large size.[64] My study area is a cattle ranch where wildlife protection is a relatively recent practice (the last 40 years or so). Presumably the really large animals might have been killed off earlier and the animals that exist now in the areas where I studied (relatively close to human activities) might not have had enough time to grow to really large sizes. But there's more to it. As we saw in Chapter 6, the anacondas in my study site may not gain any benefit in terms of reproductive output from being larger. Then again, not all benefits need to be reproductive. Larger females may have larger babies that may have a larger chance of survival. Larger females may attract more males and obtain benefits in sperm quality, resulting in higher-quality babies, so quantity is not necessarily the only part of the story, important as it may be. I will address the questions regarding maximum size in the rest of this section using the information that I have gathered.

Measurement Errors

I discussed in Chapter 2 the problems of measuring a large snake and how the measurement is less reliable if the animal struggles and if the measurement is taken by people without the proper experience.[9] This could be even more of a problem if it's a very large snake that is not muzzled and the researchers are struggling with the animal as they try to stretch it. Another complication occurs when the people doing the measuring do not really want to obtain a reliable measurement but instead are trying to impress with a tale of a very large snake, which could have been the case in some of the older records. However, I doubt that this alone can account for the large difference found.

Constraints on Large Body Size

The condition with isometry assumption (CIA) of pregnant females decreases with body size. I have argued that mass can be a problem for animals when crawling on dry land, so larger animals might be constrained by their mass. Larger females might alternatively have problems finding the energy they need to reach the breeding condition (see Chapter 6). It would not pay for the female to grow to a size where she cannot find enough food to breed. Nor would it pay for her to be so large that her mass would be unbearable (and risky). Regardless of the reasons, extrapolating the decrease in condition index, I found that a female with a snout–vent length (SVL) of 600 centimeters (approximately 676 centimeters [22.2 feet] total length [TL]) would have a CIA of 0.875, which is the cutoff below which females do not breed. In other words, if this trend holds for larger animals, a female larger than 6.7 meters TL would never reach breeding condition. Thus, there would not be any reason for the female to reach this size. However, this theoretical limit (676 centimeters TL) and the actual maximum (522 centimeters [17 feet] TL) are still far apart. At this size, 1.5 meters (5 feet) represents a difference of 70 kilograms (154.3 lbs), almost double the body mass.

A 5-meter limit for female size is consistent with the calculation that Pritchard[92] made about the maximum size of the snakes. In a review of the size at maturity and maximum size of several North American snakes, he found that the maximum size of the snakes was about 1.5 to 2.5 times the size at maturity. The minimum size of a breeding anaconda that I found was 210 centimeters (6.9 feet) SVL, so the expected maximum size of the anacondas following this pattern is 315 to 525 centimeters (10.3 to 17.2 feet) SVL. This range easily accommodates the largest snake I have found and also the maximum size predicted by the yearly reproductive value.

Explaining the Records in the Literature

How does one explain all the records of much larger animals documented in the literature? The above analysis suggests two limits on the development of larger size. One is the decrease in the CIA that sets a limit of 676 centimeters above which the females do not reach breeding condition. The other is the decrease in breeding frequency as the female grows larger, which makes it unprofitable for the female to grow beyond 5 meters. A third line of evidence, the size at which snakes reach adulthood and the phylogenetic pattern in the group, also shows a little more than 5 meters as the limit. The first two limits probably depend on the environment and are likely to be less important when the females live in more permanent water or in an area without a long dry season. Females with a steadier food supply and less danger of mortality when they are young may breed later at a larger size, so the 2.5 factor will result in a higher upper limit.

For instance, if the anaconda who lives in a river or deeper lagoon, or in the aseasonal rainforest, would not face the constraints of gravity, and hence the limitations of mass on dry land would not exist. Both the decrease in CIA and breeding frequency might well be related to a decrease in the feeding frequency or a lower supply of highly profitable prey items that occurs in the dry season. Will a very large female have enough food intake in a flooded habitat to reach breeding condition and a relatively higher reproductive frequency, despite her larger size, in more permanent water bodies? I have shown that the precipitation during the previous 2 years affects the breeding frequency of the anacondas (see Figure 6.25). So, the hypothesis that breeding frequency is higher in wetter areas is plausible. This would lead straight to the conclusion that larger size is also possible if breeding frequency does not drop with size, or if it drops less steeply.

Living in a more permanent body of water might be safer for large anacondas, too. They would not be exposed to overheating, since they do not have to crawl on dry land. The risk of being wounded during predation is relatively high for anacondas in the savanna (see Chapter 4). An anaconda that attacks prey in, or near, a river can drag it down to the water with less risk of being attacked by its relatives (e.g., capybara). Bringing the prey to an unfamiliar environment where its mobility is constrained may increase the efficiency of subduing prey. Thus, in a river habitat (or neighboring oxbows and lagoons), a large female might be able to locate her food in a safer manner, allowing her to use her energy and skills in a more efficient way since there would be less risk of injury. Having a lower risk of being injured may increase her longevity and growth since she would spend less energy healing from wounds. Longer-lived snakes, of course, will result in larger snakes.

There are other selection pressures for the female to grow larger in rivers. A 5-meter-long anaconda can kill and subdue all the native animals occurring in the savanna, so there would not be any advantage in the food intake for the female to grow larger. However, in the rivers and associated gallery forest, anacondas may find larger and more diverse prey items. Certainly if we count some of the nearly vanquished species that once were abundant, and can still be found in pristine habitats. There are large vertebrates such as peccaries (*Tayassu tajacu* and *T. pecari*), tapirs (*Tapirus terrestris*), crocodiles (*Crocodylus intermedius*), and Arrau turtles (*Podocnemis expansa*) that can be important food sources. A larger size would benefit the females by enabling them to capture these preys. Also, other less dangerous (non-social) rodents such as agoutis (*Dasyprocta* spp.) and pacas (*Agouti paca*) become available on a more regular basis than in the savanna. Animals living in the more stable but prey-unpredictable environment probably must take advantage of more diverse prey (in terms of size and species). So, larger body size would benefit females by increasing their range of potential prey. Also, in wetter places where smaller anacondas might have less predation due to the lower encounter rate with predators, the females would have less mortality at early ages. Natural selection will favor investing more energy in growth early in life, leading to larger size at first reproduction and larger total size. Thus, females in more permanent water bodies would be selected to grow large and also would have the food supply and added safety enabling them to do so.

If these assumptions are correct, I can make the following predictions:

1. Females in wetter habitats would start breeding at a larger size due to the lower encounter rate with predators in the dry season, which would make it more profitable for an "adolescent" female to keep growing in order to produce larger clutches. This seems to be supported by some anecdotal evidence. A female that was born in captivity and fed *ad libitum* throughout her life started breeding at 3.1 meters (10.2 feet) SVL (3 years of age; William Holmstrom, personal communication), a whole meter (3.3 feet; 50%!) larger than the smallest breeding animal from the field.

2. Females, being larger, would produce larger offspring. This is a consequence of their larger size but is also rather convenient because neonates may need a larger size to cope with the likely unpredictable food supply associated with the high waters.

3. Females would have a more uniform CIA across different sizes of reproductive females since there is no dry season where everybody gets skinny; alternatively, the decrease in condition with size would be less steep.

4. Females from wetter habitats may face a lower risk of injury during predation, allowing them to invest more in growth. Thus, they would suffer fewer injuries and they should live longer.

5. The breeding frequency in females may still decline with size due to scaling of metabolic processes, but it may decline less steeply than in hyperseasonal habitats since they would be able to forage throughout the year.

As a consequence of all these predictions, I expect that females might be able to reach a larger size in wetter habitats.

In fact, the model described in Figure 6.21 is very sensitive to changes in the breeding frequency, so the optimal size can easily shift upward if the snakes live in different environments where conditions are different. Furthermore, the few data that I have collected from river environments support this hypothesis. Sylvia, the female anaconda captured in the Tiputini River (Ecuadorian rainforest, Napo province), measured 522.8 centimeters (17.2 feet) TL and 459.8 centimeters (15.1 feet) SVL. This is the second largest individual in terms of SVL in the 28 years I have been studying them (Mirna, who was 477 centimeters in SVL, had a stumpy tail). The one animal caught in the rainforest was larger than almost 1,000 of the individuals caught in the seasonal savanna. Sylvia had only 13 scars, far less than expected for an animal of her size (see Chapter 4), supporting the hypothesis that animals in more aseasonal habitats face less risk of wounds from their prey. Also, among the few males I have caught in a river, 4 are among the top 95% percentile of the 429 I have collected (Figure 7.1). Anecdoticaly, a 372-cm long male was in Formoso River (Cerrado, Brazil, Juliana Terra Personal Communication) and a 4.5-meter long males was caught in the Peruvian Amazon (Renata Leite Pitman, personal communcation). While two data points does not constitute conclusive proof, they do point in the same direction than my former analysis. The literature also shows that most of the very large snakes come from rivers.[86]

Let me offer a few anecdotes in which I believe the observers had seen a very large snake. Mayer was a guide in the Tiputini field station in Amazonian Ecuador (where Sylvia was caught) and he had deep knowledge of the forest. Having worked with scientists for years, he was aware of the importance of accurate information and how hard scientists pursue facts. He reported seeing an anaconda with a head about 20 centimeters (7.9 inches) long (as he showed me with his hands). He also said that coiled up on the riverbank, she was about 50 centimeters (1.6 feet) tall. This head size corresponds with an anaconda a bit shy of 7 meters (23 feet) and 25 centimeters (9.8 inches) across. Coiled over itself, it would be close to the reported height. Not having seen it stretched out Mayer did not venture to guess its length, but this is an internally consistent story that suggests he had seen a very large anaconda indeed.

Also, an old worker on one of the ranches where I did my research told me that he had spent some time during his youth killing anacondas to sell the skin, way back in the 1970s or so. He told me that he had once killed a very large anaconda, a *madre de agua*. "It was an enchanted snake," he said. "It was so large that it was growing horns." I have already explained how keloids on the head of snakes may give the impression of antlers developing, and this is likely what the old worker saw. He explained that the snake was as big across as a barrel, pointing to a plastic barrel that was 61 centimeters (2 feet) across, and said that the ventral scales were as wide as his palm. When I asked him about the total length, he told me that he didn't know because after he killed it he realized what he had done and, fearful that he had killed an enchanted animal, he left without collecting the skin. Again, this is an internally consistent story.

Finally, conversations with Guarao Indians in the Orinoco Delta resulted in another internally consistent report of a very large snake, far larger than the ones I have found in the hyper-seasonal savanna. Common to these sightings is that they have occurred in permanently flooded places, removed from human activities.

It is important to point out that human activities would preclude the conditions that favor the development of very large size in a snake. Larger prey items (tapirs, capybara, peccaries, crocodiles, and Arrau turtles) have decreased in number due to human activities and hunting; indeed, most of these species are endangered. Human presence also would prevent the anacondas from growing to large sizes due to the propensity of people to kill snakes. Thus, truly large snakes are probably restricted to permanent waters with little or no human presence.

What About Old Extinct Giants?

In 2010 the scientific community was shaken by the discovery of the fossilized remains of a snake of formidable proportions in eastern Colombia: *Titanoboa cerrejonensis*.[88] Unlike other giant reptiles that have impressive fossilized skulls, snake fossils are not necessarily that showy. You may find a 2-meter-long (6.6-foot-long) skull of a crocodile whose size impresses even the layperson. Snakes' heads, on the other hand, have too many moving joints; even a very large snake doesn't have a large piece of bone. Certainly there are many vertebrae and many ribs, but they do not show the magnificence of the animal unless one does the mathematical calculations to determine the size. The vertebrae of Titanoboa, at more than 10.2 centimeters (4 inches) across, did not seem to belong to a formidably large animal. In fact, for several years they were thought to belong to some unimpressive species of crocodile. It was only on closer examination that they were determined to belong to a giant snake. By way of comparison, vertebrae of large anacondas I have found are less than 2.5 centimeters (1 inch) across. Eventually the size of Titanoboa was calculated to be as large as 12.01 meters (39.4 feet) SVL and 12.82 meters (42 feet) TL.

Titanoboa lived 58 to 60 million years ago, shortly after the demise of the dinosaurs. It was an aquatic boa that lived in a large swamp that likely covered a lot of what is now eastern Colombia and western Venezuela. This swamp must have been a very interesting place not only for the giant snake but for the other fauna that shared the swamp with her. There would have been a vast variety of fishes, but also a few crocodiles, and magnificently large turtles up to 3 meters (9.8 feet) in length. Later on, there were capybara-like mammals as large as a horse, and even a small elephant about the size of a bull occupied the area, but these animals appeared in the fossil record much later. Nevertheless, it was a rich tropical habitat whose real diversity is bound to be underrepresented in the fossil record. We can know for sure, though, that

Titanoboa had a rich menu of amazing vertebrates at its reach. Whenever I think of this, I get this strong craving to have a time machine to go there and check out this fascinating world!

Phylogenetically Titanoboa is closely related to anacondas. Being large aquatic constrictors, anacondas are likely the closest live species of snake in both ecological niche and phylogeny for us to have an idea of what Titanoboa was like. It isn't often that we have such a good current model of an organism that resembles an extinct one so that we can make good guesses about the life of the extinct giant. In contrast, ostriches are poor surrogates for *Tyrannosaurus rex*! Using what I have learned of anacondas, I am prepared to make guesses on what Titanoboa was like. But I need to use caution because, despite how similar Titanoboa may be to a current anaconda, the similarities only go so far. For instance, Titanoboa was not necessarily a scaled-up anaconda. It was a closely related organism, perhaps very similar in many ways, but it lived in its own habitat, had its own selection pressures, and dealt with a different assortment of predators and prey than today's anacondas do.

What Did It Look Like?

Like most herpetologists, I was in awe when I first read of the discovery. After I finished wondering how magnificent a beast it must have been and daydreaming what it would be like to have a personal encounter with one of them, I had a sudden realization and nearly fell out of my chair: It had a stumpy tail! In my study site I have caught anacondas 5 meters (16.4 feet) long that have tails of more than 50 centimeters (1.6 feet). Titanoboa was double that length yet its tail was only 82 centimeters (2.7 feet)! At 12 meters (39.4 feet) SVL, one would expect a tail of no less than 1.47 meters (4.8 feet). Based on the raw data (Figure 7.2), it seems like Titanoboa's tail was seriously stumpy. I analyzed the dataset provided by the authors to recalculate the size of Titanoboa, thinking there might have been an error. The data were scrambled and I was unable to repeat the calculation of their size. If Titanoboa's tail was not 82 centimeters (2.7 feet) but 1.3 meters (4.3 feet), then she was not 12.82 meters (42 feet) TL but rather 13.4 (44 feet)! I eventually obtained the proper dataset from the authors and confirmed that their calculation was correct.

Extrapolating the size of an anaconda tail using its body as predictor, not surprisingly, males have longer tails than females, but even using the regression for female anacondas would still predict a much longer tail than Titanoboa had. However, looking at the relationship between tail length and body size in anacondas, one can calculate the relative length of the tail associated with body size. Figure 7.3 shows that anacondas' tails become shorter, compared to the length of the body, as they grow larger. If an anaconda would grow to the size of Titanoboa, its tail would be on par with the length of the tail of the extinct giant. These calculations humbled me and gave me a new appreciation of how similar Titanoboa and anacondas really are—not only phylogenetically, in habitat use and morphology, but possibly even in their allometric growth!

Of course, the first thing we think about would be the size of Titanoboa. It's one thing to say that it was 12.8 meters long but quite another to picture what that would look like! Using regressions from anaconda size, she was estimated to weigh 875 kilograms (1,929 lbs) and to measure about 46 centimeters (1.5 feet) across. Now I can do a finer calculation, since the regression used originally included males that were skinnier. Among boas, larger females are the rule, so it is reasonable to assume that the largest Titanoboa was a female. A non-pregnant Titanoboa would have weighed 1,232 kilograms (2,716 lbs). Using regressions of pregnant females, we see that a pregnant Titanoboa would weigh a whopping 1,465 kilograms (3,230 lbs) and would measure 76 centimeters (2.5 feet) across. This exercise allows us to get a bit better acquainted with what it might have looked like.

How About Reproduction?

However, picturing what Titanoboa looked like just didn't feel like enough for me. I wanted to explore what its life history was like. I can use some of my data and what I know about anacondas to make educated guesses about some aspects of Titanoboa's biology. For instance, the paleontological findings point out that Titanoboa reached a size as large as 1,201 cm (39.4 feet) SVL. But what was the size of most Titanoboas? When did they start to breed? How big were their babies? How many babies did they have? Since the data on snakes and size at first reproduction work so well in predicting this information in anacondas, I would use it to make a guess about when Titanoboa started to breed. I will assume that the longest reported length of Titanoboa is the maximum size it reached (but I mention some caveats later in the chapter). Based on information from current snakes, they grow to 1.5

to 2.5 times the length at which they start breeding.[87] In the particular case of anacondas, the 2.5 figure is closest to what I found. So, using this figure, if Titanoboa reached 1,201 centimeters (39.4 feet), it must have started to breed at 480 centimeters (15.7 feet) SVL, the maximum size reached by current anacondas in the hyper-seasonal savanna (Figure 7.4).

An anaconda of the size of a new Titanoboa mother would weigh 95 kilograms (200 lbs). At this size she would have 88 babies averaging 323 grams (0.71 lbs). If a modern-day anaconda were to grow to the size of Titanoboa, extrapolating the clutch size and other parameters, because the investment per offspring decreases with the size of the female, she would invest 0.04% of her body mass in every baby. She would produce 1,321 babies weighing 392 grams (0.86 lbs) each. She would invest 505 kilograms (1,113.3 lbs) in babies, amounting to a whopping 55% of her non-pregnant body mass.

Now, these numbers are cool and impressive, but they are not very responsible. Basically, I have scaled up the figures of modern-day anacondas to the size of Titanoboa. However, was Titanoboa a scaled-up anaconda? Was its ecology just like that of a present-day anaconda, just one that happened to grow more? This is highly unlikely. Titanoboa lived in a different time, surrounded by different organisms, predators, and evolutionary selection pressures. Undoubtedly, its ecology and selection pressures responded to a different scenario.

For instance, Titanoboa's world comprised large predators, several species of crocodiles, predatory fishes that abounded on the waters, and likely some other terrestrial ones. It was just after the demise of the dinosaurs, when many organisms were a lot larger than now. If a Titanoboa, had babies of the size of today's anaconda (393 grams [0.87 lbs], 90 centimeters [3 feet] SVL), this baby would have faced a large number of very large predators. Current anacondas face lots of predators, too, but after a couple of years they outgrow many of them. and after a few more years they are literally predator-free. It was likely not that different in Titanoboa's times. If Titanoboas had been born the size of current anacondas, their babies would have to endure lots of predation for a very long time, which probably would have prevented them from reaching adulthood and reproducing. Also, modern-day anacondas have a variety of smaller prey that the babies can feed on. Likely these smaller prey were not that common in Titanoboa's time. We have a similar scenario as if anacondas that bore babies the size of water pythons discussed earlier. The minimal size at birth corresponds to the specific selection pressures. So, a Titanoboa the size of a current-day baby anaconda, scaled up as it may be, would have faced very long odds, both for finding small food items it could consume and also for the odds of being preyed on during its long time of vulnerability.

I will assume that Titanoboa was the ecological equivalent of a current anaconda (the largest aquatic snake on the block), an assumption that is reasonably well justified. A young female Titanoboa ready to breed likely faced the same decisions as current-day anacondas. She had to be able to produce babies that were ready to survive in terms of both avoiding predation and finding food. Natural selection would have shaped the natural history of female Titanoboas so they would breed as soon as they could produce a viable clutch both to increase their reproductive output and to secure some offspring in case they were preyed on before the next breeding opportunity. I will assume that their babies had to be the same proportional size as the babies of current-day anacondas. If a Titanoboa's babies represented 4% of her mass, they would weigh 3.9 kilograms (8.6 lbs) at birth and measure 1.8 meters (5.9 feet) SVL. I can use similar reasoning to estimate what the babies would be like when she is full size. The relative investment per offspring declines as the animal gets larger; in anacondas, it drops to 0.56% of the female's mass. If the mother's mass 1,232 kilograms (2,716 lbs), a Titanoboa neonate would had weighed 6.9 kilograms (15.2 lbs) and measured 2.16 meters (7 feet) SVL at birth (Figure 7.5).

Meal Size

Of course, an animal as large as Titanoboa had to consume very large animals, and there were a whole assortment of crocodiles, turtles, and fishes at the time that would have done nicely. Regardless, we can use our knowledge on anacondas to make an educated guess on their prey size. Average prey size of neonates is about 32% of their body weight. So, neonatal prey must have averaged 2.2 kilograms (4.9 lbs). My data on anacondas tell me that anacondas can go for prey as large as 146% of their body weight. Extrapolating we can calculate the size of an average female, an average mass for Titanoboa was 1,232 kilograms (2,716 lbs). If a Titanoboa were to attack prey 146% of its body size, this prey would be 1,799 kilograms (3,966 lbs)— the size of a small SUV! Titanoboa would have been able to swallow prey as large as most current terrestrial vertebrates, with the exception of elephants and rhinos.

Now, let's not succumb to the charm of the amazing and pay a little more attention to the cool. The average prey size of anacondas is 30% of the female's body mass. For Titanoboa this would be 370 kilograms (815.7 lbs), the size of a modern heifer. Like any snake that feeds on very large prey, Titanoboa was likely able to shut down its metabolism when it was not foraging. Alternating between feast and famine is common among the majority of the snakes, and certainly among boas, so there might have been times when Titanoboa went for long periods without a meal. Letting their digestive system atrophy saves energy when digestive tissue is not being used. This may be the reason that anacondas drop smaller prey from their diet as they grow larger: It just doesn't pay for a very large snake to restart its large metabolic machinery for a prey that is not big enough. Looking at the average prey size of female anacondas larger than 3.3 meters (10.8 feet), meals represent 41% of the female body mass. So, an average human being weighing 70 kilograms (165 lbs) would weigh only 5.7% of Titanoboa's mass. Anacondas hardly ever take a prey this small, so there is a very good chance that Titanoboa was so large that it wouldn't have been too interested in eating a person. This piece of information broke the heart of the movie producer!

How Big Again?

Last, let me leave you with a reflection about statistics and chance. Paleontologists know all too well how difficult it is to find fossils. First, the odds that an organism would fossilize are long. If the animal is eaten or decomposed, it will leave no fossil. The great majority of organisms do not fossilize. Second, the fossil needs to be found by someone who knows it is a fossil, not a weird rock, and identified properly. Considering all this, we are lucky to have any fossils at all.

Figure 7.6a shows the distribution of female anacondas from my study. We need to wonder if the Titanoboa fossils that have been found came from one of the very largest snakes (like the ones on the right-hand side of the figure) or from just an average-sized (relatively speaking!) individual. All the calculations, and what the scientific community accepts as the size of the largest snake that ever lived, are based on the fossils found. These provide the only rock-solid evidence that it existed (pun intended). But what are the odds that the fossils represent the very largest Titanoboa? What are the odds that the very largest Titanoboa actually fossilized, and then 60 million years later

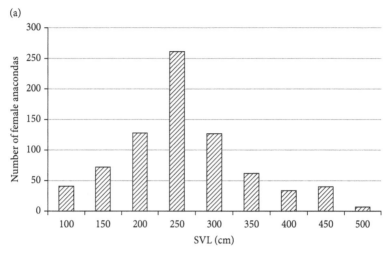

(a)

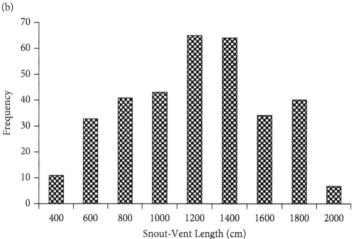

(b)

Figure 7.6 (a) Distribution frequency of female anacondas in the llanos. The most common individual is one of about average size. (b) Hypothetical size distribution of Titanoboa assuming that the fossils found belong to average-sized individuals and that Titanoboa distribution follows the same pattern as that of current-day anacondas. If the Titanoboa fossils found represent only the "most common" individuals, its true dimensions must have been truly mind-boggling!

was found by someone who recognized it for a fossil? There is a good chance that the largest snake that ever lived will never be found, because it represents only one snake; the odds are against finding even the largest few.

However, in the Cerrejón mines where Titanoboa was found, they have found several individuals of about the same size. This opens the window to a fantastic notion. What if the Titanoboa that was found was not the largest at all? What if the 12-meter-long Titanoboa found might have been just your run-of-the-mill, regular-size, impress-nobody Titanoboa? Figure 7.6b also shows a hypothetical distribution of what Titanoboa populations would have been like, if they followed the same pattern than current-day anacondas and the one that was found was an average-sized individual. If this is the case, it is not impossible that the largest snake that ever lived was far larger than the Titanoboa that has been found—a theoretical maximum shy of 20 meters (65.6 feet) in SVL, 22 meters (72.2 feet) in total length and 5,091 kilograms (11200 lbs)! But of course, in science we stick to what we know for sure.

8

"Hey, It's Me, It's Me!" Courtship and Confusion in the Male Reproductive Strategy

Una excursión de pesca en los llanos encontraron un "rollo de culebras" que era formado por una hembra y muchos machos enroscados a su alrededor. Unos peones Llaneros que les acompañaban dieron muerte a la hembra para quitarle la piel; a tal fin la ataron por la cola y arrastraron por la sabana un largo trecho. Una vez hecho esto procedieron a desollarla. Por la mañana siguiente se encontraron con que todo el rastro dejado por la culebra estaba tejido y surcado por muchísimas huellas de otras culebras. Evidentemente las mismas habían sido atraídas por el almizcle que había esparcido la hembra por el camino.

(A fishing expedition found a "roll of snakes" that was formed by one female and many males coiled around her. Some llanero peasants that accompanied them killed the female to take the skin. To this end they dragged her by the tail for a long stretch. Once they had done this, they dressed it. Next morning, they found that the entire trail left by the snake was weaved and crisscrossed by very many tracks of other snakes. Evidently these had been attracted by the musk that had been left by the female on the trail.)

Excerpt from *Fauna Legendaria de Venezuela*[12] (Translated by JAR)

Becoming a Biologist

Being a biologist has been in my mind since I was so little that I cannot remember. One of my uncles, Eliso, an electrician, says that when I was very little, about 4 years old or so, he asked me what I wanted to do when I grew up. I answered that I wanted to be a biologist and study marine and aquatic organisms. To antagonize me he told me in a commanding tone, "No, you will be an electrician!" As he tells the story I replied, unfazed, "Fine! I will study the roaches and creepy crawlies of the electrical outlets." This shows not only my early decision to be a biologist but also my stubborn determination to do it despite the circumstances, which is a prerequisite for anyone who wants to succeed in this field.

Years later the same uncle gave my brother and me two rabbits. It was simply a prank on my mother (his sister-in-law) since he knew that once we saw the bunnies she wouldn't be able to talk us out of keeping them. This was not an easy task in a two-bedroom apartment in the middle of a big city. Since my brother is older, he beat me to choosing a rabbit and he chose the male, a sweet and tame black animal. This left me with the female, a calico, who was wild and feisty one and would not let anyone near her. As I was so little, I don't recall many of the details. I do remember reaching under the living-room table with an extended hand to establish some sort of contact with the wild lagomorph, trying to convey to her that I loved her and meant her no harm. I wish I could tell how it happened, but a few days later my mother was surprised when she was in the kitchen and caught me out of the corner of her eyes carrying a rabbit in my arms like it was a baby. She did a double take because the rabbit was not black, the tame one. Then she realized that I had managed to tame the wild rabbit. Somehow, not being fully able to speak, I was still able convey to this rabbit that she had nothing to fear from me. Not only that, but from that day on the rabbit understood that people were not dangerous and stopped being skittish altogether!

When I was finishing high school, the time came to make a decision about what career I needed to study. Since I was good in math, chemistry, and physics, there was a lot of pressure on me to study engineering—"the better to get a good-paying job," my father would advise. Yet, despite the influential figures and professional counselors from high school who pushed me in that direction, I still resisted. Any talk of studying biology was met with a comment like "To do what? You won't be able to get any decent job. All you can do is teach with that career." The funny thing is that both my parents were professors at the university, two of my grandparents had been teachers, and all my aunts were teachers too. Even Eliso, the electrician, ended up getting a master's degree in education and became a teacher as well! So, I felt that teaching was something that kind of ran in my family. Yet I felt the pressure to go into engineering. I went as far as taking a qualifying exam at the university's school of engineering, which I passed, and I was accepted. This was no small potatoes because university education in Venezuela is free, so admission is very competitive, certainly at Universidad Central de Venezuela (the main and most prestigious university). However, when the day came to make a decision, I envisioned myself with a hard hat working on a construction site and I was just not very taken with the idea. Then, I recalled being under the living-room table with the bunny on my chest, holding her nose right up to mine and looking deep into her eyes. Then I knew that I was a biologist, regardless of how difficult it would be to get a job. I turned down the position at the engineering school and enrolled in the biology program. I have never looked back!

When I started my college education, I was so taken with snakes that I couldn't wait to take the upper-level courses where one gets to go to the field and start learning about specific organisms. I went to the local zoo and sought learning on my own. I went to the Parque Del Este, the same zoo where years earlier I had been fascinated watching the anacondas in their open-roofed pit, and looked up the person in charge of the reptile house. It so happened that the herpetarium had a very low budget and did not have a lot of personnel. A manager in his 60s led a handful of workers who took care of the facility, like custodians, but wouldn't come near snakes with a 10-foot pole. They took care of the mice colony and handled upkeep of the facility, but everything that involved working with snakes was for the "old man" to do. The manager, Jesús Carrasquero, had some health issues and it was hard for him to do all the work needed to clean the cages, feed the snakes, and assist the animals in what they needed. He was delighted to have an enthusiastic 16-year-old ready to learn and do anything he asked me to do. We had a perfect symbiosis. He taught me everything I wanted to learn about snakes, how to handle them and how to take care of them. I was an eager apprentice ready to absorb any knowledge or information he told me. In exchange, he got all the work done by me. Once he saw that I had learned the ropes, I was running the reptile house pretty much by myself. In retrospective one has to question the wisdom of letting a 16-year-old, motivated as he may be, run a reptile house full of all kinds of venomous snakes, including rattlesnakes, fer-de-lances (Terciopelos or Mapanares), and a few others. I am so glad I grew up in Venezuela; try that one in a zoo in the United States! Granted, at 16 I looked like I was 20-plus, but he never inquired about my age and I never volunteered it!

At any rate, working at the reptile house gave me the chance to learn a lot about snakes, and my certainty about my career path became clearer: no marine biology, no roaches from electrical outlets, but snakes. However, I needed to choose a topic for my undergraduate thesis (an undergraduate degree in Venezuela requires a master's-like research project). I wanted to do it on snakes, but I couldn't find funding or a mentor who knew anything about them. Gerardo Cordero, a professor from Universidad Central de Venezuela, adopted me as a student, but his specialty was small- and medium-size mammals. His interest in vertebrate biology was broad enough, though, that he would entertain having me as a student. Under his direction we started going to his field site in Rio Negro, located in a spot of wet forest in the northern part of Venezuela in a region called Barlovento. While Gerardo did his work with mice, I was out on my own looking for snakes. Not having funding for drift nets or any other sort of formal sampling, or having knowledge of them really, I counted on opportunistic captures. Sometimes I did great, other times not so good. As time passed, I realized that I was learning a lot of field biology, mammal work, and bird work as I assisted other scientist on site in my spare time. I was learning a lot about snakes, too, but I realized I was not going to be able to get enough data to write my undergraduate thesis.

This experience taught me a lot, including how deep the fear of snakes runs in some local communities. In the field station, one of the night guards, Tata, used to make jokes all the time, often making fun of the scientists. Students, of course, were an especially easy target for him (Venezuelans have a peculiar sense of humor). Often in the evenings we'd listen to Tata taking pot shots and making rude jokes about everybody present. We spent quite a bit of time listening to his jokes in the dining area, which doubled as the sleeping room when we all hung our hammocks at night. After a few months of working there, someone noted that Tata had not given me any nickname and I was never the butt of his jokes. This was unusual because everybody else had one or two nicknames, and they were always very derisive. When questioned about it Tata gave me a nickname on the spot, Canagüita, but quickly indicated to me that it was good. I knew that it meant "little Canagüa" but didn't know why he gave me that name or what Canagüa meant. Tata later told me that Canagüa was a powerful warlock who lived in the nearby town of Birongo. He was very well respected and very feared. The area used to be cocoa and coffee plantations, and the inhabitants of Barlovento are mostly descendants of the African slaves who worked on the plantation. Their original African religions have merged with Christianity seamlessly, not unlike what happened in the Caribbean islands with voodoo, Santería, and other religious practices. So, witchcraft in the area is rather important, and witches and warlocks are significant players in everyone's life. Calling me "Canagüita" was meant not as an insult but as a compliment.

Tata had seen me go in the forest at night armed with a flashlight, a repurposed trashcan, and a homemade snake hook I'd improvised from a mop handle. I was so poor, I didn't even have a decent headlamp for the job. Local people have true respect for the forest at night. Nobody goes there without at least a partner and a couple of shotguns. There are jaguars in the forest, pumas, and of course snakes. My going into the forest unarmed and alone was enough of a feat of bravery by itself, but returning after a few hours with a load of snakes, including the fer-de-lance, could only be explained by witchcraft. The respect that the local people have for the *boca frías* (cold mouths), as they call fer-de-lances in the area, is not the healthy respect that one should have for a venomous snake; rather, it is a superstitious fear of a superior force. *Boca frías* are not considered dangerous; they are viewed as evil. For a kid like me to go into the forest, catch them, subdue them, and manipulate them with my hands was beyond normal. I had to have some secret powers, and Tata didn't want to get on my bad side!

I eventually stopped working in Rio Negro and ended up settling for iguanas, the closest I could get to snakes for a study subject and still obtain funding for it. I studied iguanas for a few years and actually fell in love with them. In fact, when I came to the United States to study for my Ph.D., my goal was to study iguanas. I had made some interesting discoveries in the behavior of green iguanas working with my professor of animal behavior, Luis Levin. It was not part of my thesis work but something that I discovered during my thesis work, and I became fascinated by it. I contacted Gordon Burghardt from the University of Tennessee, who had done some early work with iguanas that had inspired my work. Gordon was very interested in the study I had done. His acceptance letter to my request was so enthusiastic that it made it easy for me to get funding to enroll in the ethology program at the University of Tennessee for my Ph.D. So, I have been working with reptiles and amphibians for the better part of three decades and most people would rightly call me a herpetologist. I have done a lot of studies of population dynamics and demography, and I could be called a demographer or population biologist. I have been doing ecological research in the tropics all my life, so people could call me a tropical ecologist. However, I see myself as an ethologist. In the rest of the chapter the ethological perspective takes front and center.

The Disparity Between the Sexes

One of the first things that struck me when I started studying anacondas was the very strong sexual size dimorphism (Table 8.1). Hidden behind the proverbial large size of females is the contrasting small size of male anacondas. Clearly, "small" and "big" are relative! With an average of 2.3 meters (7.5 feet) in length and 7 kilograms (154.3 lbs) in mass, male anacondas qualify easily as very large snakes. However, they represent only a fraction, about one fifth, of the size (mass) of the average adult female. The difference in size between males and females is so dramatic that a female could be 40 times more massive than a male mating with her. As we have seen throughout the book, this size difference influences just about every aspect of anacondas' biology. It produces difference in prey size, prey choices, mortality, and energy budget, to mention just a few. Considering all these differences, we must ask: Why are male anacondas so much smaller than females? In the prior chapter, we discussed the selection pressures leading to large size in females. In this chapter, I will attempt to show, to the best of my abilities, what selection pressures lead to the dramatically smaller size in males.

The Advantages of Being Small

There are a number of advantages for animals to be small. Small animals have lower metabolic costs since they need to maintain a small body that requires less food. Clearly, foraging is a potentially dangerous activity. Animals that need less food have potentially higher survival than those that have a large body to maintain. This is particularly important during droughts or times of food shortages. Also, male anacondas feed on smaller prey that does not

Table 8.1 Measurements of adult anacondas found in breeding aggregations.[18]

	Total length (cm)				Snout–vent length (cm)				Mass (kg)			
	Mean	SD	Min.	Max.	Mean	SD	Min.	Max.	Mean	SD	Min.	Max.
Females (N = 48)	370.4	70.6	242.7	517.3	326.2	65.9	210.7	477	32.60	18.59	9.25	82.50
Males (N = 177)	263.2	28.3	188.3	333.7	225.9	24.7	159.3	293.7	6.96	2.07	2.45	14.30

Reproduced with permission, Eagle Mountain Publishing, LC.

necessarily provide a lot of energy. Because of this they have significantly fewer scars and lower mortality than animals that attain large sizes (see Chapter 4).

A small body is also an asset because it lowers the cost of locomotion. To haul around a large body incurs larger energetic expenditures than if the body is small. This is particularly important for males, which move much more than females in the mating season to track receptive females for mating (see Chapter 5). If a male grows too large, he would incur a higher cost of locomotion. This is especially true if he has to travel over dry land in the mating season. Being small also offers other benefits such as being inconspicuous, not being a very coveted meal, and hiding easily from potential predators.

Of course, there are also some disadvantages of being small. Small animals cannot fight off predators as well as large ones, and their gape limits the kind and size of prey they can eat. Furthermore, larger males may produce more sperm simply because their gonads are larger. Larger testes can produce a larger amount of sperm, which would give them an advantage in siring more offspring, particularly if the female mates with several males. Last but not least, larger males can compete physically for the females by excluding other males through physical struggles.

When Males Are Larger

When I was in Tennessee doing my Ph.D. I had the opportunity to visit a project going on at the Knoxville Zoo with silverback gorillas. Males fight ferociously for control of the females and have no tolerance for other males. I was invited behind the scenes and went in with one of the keepers, a female. I was excited to see one of these massive gorillas up close. When I first saw him, I looked straight into his eyes, fascinated by the wonderful animal. But, by drawing myself up to my full height, sporting a significant beard, and looking into the eyes of a silverback gorilla, I was not conveying my true admiration and fascination for the animal; rather, to him this represented a challenging position and gesture. My 83-kilogram (183-lb) human body was challenging his 200-plus-kilogram (440-plus-lb) gorilla might. The enraged gorilla charged at me full speed. I froze during the split second it took the gorilla to travel the 4.6 meters (15 feet) between his resting spot and the edge of his cage where I had challenged him. He slammed against the thick Plexiglas barrier with his two huge fists and elbows with all the power of his short run.

Then I remembered that it is not a good idea to make eye contact with a large primate and looked down, adopting the submissive position I should have had all along. I am glad the Plexiglas was so thick!

Throughout the animal kingdom, there are few rules that hold across a broad diversity of taxa. One of them is the rule that when males compete physically over females, males tend to develop larger sizes and/or larger weapons. This rule is very prevalent among mammals, and thus in many mammals males are larger than females (as in the lowland gorilla). The difference in size between males and females is called sexual size dimorphism. A lot of people tend to believe that male-biased sexual size dimorphism is the trend in the animal kingdom, but this is due to our mammalian bias. Now that other vertebrates are better known and better studied, it is clear that females are, in general, larger than males. In fact, since blue whales have female-biased sexual size dimorphism, and they are the largest animals that have ever lived, it is more than likely that the largest animal the world has ever seen, living or not, is (or was) indeed a female.

If males compete over females, there are some males with larger reproductive output than other males. When the competition is physical, then there is selection pressure for animals to be larger. This is the reason that male gorillas, silverbacks, are so much larger than females. Larger males mate with several females, consigning other males to bachelorhood. Often a small difference among males results in a large difference in reproductive success. Thus, only the very large males will produce any offspring and most other males are unable to breed, even if they are quite big themselves.

We can see a mundane example of this by looking at guitar players. You may stroll into a bar in Nashville any weekend and find any assortment of bands playing live music. If you go to the right place you can find a band whose guitar player is outstanding, on a par with guitar players who are raking in millions of dollars in well-known bands. Yet, our friend from the Nashville bar will probably spend his whole life playing for the drinks and doing odd jobs on weekdays just to make a living. Simply put, a guitar player who is 90% as good as Jimi Hendrix does not make proportionally the same kind of money or is proportionally as well known, even though he is a mighty good guitar player. This is the kind of asymmetry of reproductive success we see in males where competition for access to females is very strong. Very good males are not proportionally successful in leaving offspring; rather, one, or few, monopolize all, or most of, the reproductive output.

Among snakes there is an almost universal relationship of male-biased sexual size dimorphism when males compete physically over females. Larger males outcompete, displace, or deter smaller males from coming near the females and thus increase their chance of mating. When males fight physically over females, larger males sire more offspring and natural selection favors larger size in males. One of the points of contention with this general rule is that some argue that when males compete physically over females, they grow larger than they would if there were not physical competition—however, this does not necessarily lead to any hard prediction about whether males are larger *than* females. Selection pressures for females being large, say higher fecundity, may still be stronger than, or equal to, selection pressures for large size in males. However, if the males mate with several large females, with enhanced fecundity, they would benefit from that enhanced fecundity. Thus, the intensity of selection would be stronger in males, and they would be expected to develop larger size to benefit them from (1) the larger number of offspring they sire and (2) the higher fecundity of the large females they mate with. Notice, though, that for this to be true, males need to mate, and sire babies, with *several* females. There must be a polygynous mating system.

My first forays into studying anaconda biology showed me the strong sexual size dimorphism they had. With these observations, and given that the majority of species show female-biased sexual size dimorphism, it was pretty straightforward for me to predict that males do not fight physically over females. If they did, males would obviously be a lot larger than they are. Little did I know then that this precise observation was going to give me one of the most relevant insights in my study.

Amorous Anacondas: Disentangling Anacondas' Mating System

Studying the mating system of snakes has not been free of glitches. Original work on the mating system proposed the existence of several kinds of polygyny depending on the availability of females, their predictability, and the length of the snake mating season.[45,89] These studies somewhat arbitrarily ruled out the possibility of polyandry among snakes for phylogenetic reasons, but no evidence, phylogenetic or otherwise, was provided supporting that assertion.[90]

Polygyny among all snakes was the status quo back in 1992 when I started my study. As I started learning about the biology of anacondas, it became apparent that it was not the case in anacondas. My data on anacondas show several males courting, and probably mating with, one female. It also showed females mating repeatedly, likely with different males in the breeding aggregation. However, I found no evidence of philandering among males. Males that found a female stood with her as long as she was receptive without looking for other females to mate with. What my data were telling me was that the anacondas' mating system was polyandry, not polygyny.

This discovery led me to think that anacondas were exceptional among snakes in practicing polyandry. However, reviewing the literature I saw that it was not necessarily the case. In fact, the case for polygyny among snakes was weak at best. Polygyny was, in most cases, being assumed or deduced from captivity trials (that were set up under a polygyny assumption), but it was very seldom documented in the field. Furthermore, field studies have been most reluctant to acknowledge the existence of mating systems other than polygyny among snakes, sometimes even despite tremendous evidence to the contrary. This happened to the point that the mating of garter snakes in multi-male aggregations, on occasion with hundreds of males courting one female, somehow was viewed as polygyny!

FSLF Seeks Chivalrous Harem of Males

If a female anaconda were to write a classified advertisement, it would read something like this: "Fat single large female seeks chivalrous harem of males to court her for weeks on end with the bargain chance of getting eaten at the end of the mating, to help boast offspring survival." You wouldn't reply to this ad? Then you're not thinking like a male anaconda, because the spirit of the ad is pretty much what the mating of anacondas is all about. This stresses the importance of our differences and how difficult it is to gain a true understanding of the life of other species.

While looking for snakes in the dry season we often find mating aggregations of anacondas. These are quite remarkable sights that can pop up just about anywhere in the swamp, often around the edges of the water or near shallow areas where females may bask if they please. Normally several males coil around the female in shallow water, forming a mass that usually breaks the surface through the aquatic vegetation. Occasionally, the female is at the water's edge or partially buried in mud and drying vegetation. Large females are often found in very shallow water or even on dry land (Figure 8.1).

Having being a fireman for almost a decade and having studied anacondas for more than two, I have heard all the bad jokes associated with "big hoses" or "large snakes," and I will spare the reader from any allusion to orgies, gang bangs, sexual depravity, or what have you. The best way I can describe the sight is with this image: Imagine a spaghetti bowl 3 meters (10 feet) across. Now imagine those noodles moving in very, very slow motion. That is what a breeding aggregation looks like (Figure 8.2).

Catching the breeding ball and collecting data on all the animals proved to be pretty challenging. For starters, you were risking a bite. Since the snakes are all together and we normally there are not enough people to collect them, there is no way to catch them all at once. The only solution is to start picking them up slowly and gently, starting with the males, hoping that the female does not realize what is a foot and takes off. If we succeeded in collecting the males without anyone freaking out, it was a good day. We could keep track on who was in the mating position, if there was an actual copulation at the time, and all the information. But if the female realized there was a problem, I had to dive on top of them, hoping to pin them down as much as I could while my helpers helped pull the animals away one by one, trying not to lose any animals. This was when we often got several bites. I take no pride on those bites, as I explained earlier, but it was just difficult to avoid them while collecting all the animals and keeping tabs of who was mating with the female. But because the smaller males were the ones doing the biting, the bites were of relatively little consequence.

Who Is Invited to the Ball?

One of the first questions that come to mind when finding a large aggregation of anacondas, since they are not easy to find, is: How do they find each other? How do males know where a breeding female is, and how are they so sure it is breeding, as a large female sometimes make a meal of the much smaller males (Figure 8.3)?[34,35,91] All snakes have a superb sense of smell, which is further aided by the use of their tongue and the organ of Jacobson, or vomeronasal organ. The Jacobson organ is a chemosensory organ that is similar to the nostril-based olfaction. Snakes' characteristic tongue flicking provides information to the Jacobson organ, located on the roof of their mouth. The tips of their tongues pick up chemicals in the air and bring them to the Jacobson organ, where very sensitive chemical receptors analyze the information and send it to the brain. This gives snakes an extremely sophisticated sense of smell. So, the natural answer to the question of how males find their females is through smell.

However, despite their good sense of smell, finding the female may not be easy. The pheromones released by the female to advertise her readiness to mate may not be easy to detect. Although no specific studies exist with anacondas, other species of snakes have been found to produce a lipid-soluble chemical that is supposed to be responsible for attracting males.[76,78,92] The problem with these pheromones is that they are not very volatile; in other words, they don't travel through the air very readily. Because lipids are hydrophobic, these pheromones do not dissolve in water easily. In other words, the males have to physically run into the scent trail on the ground in order to find the female. This may explain the enhanced mobility of males during the mating season: They are trying to find the trails of receptive females.

Aquatic snakes are supposed to have limited abilities to leave and follow scent trails, which may be difficult to leave in the water.[70] However, since we do not, and cannot, smell water-borne scents, we are perhaps handicapped in understanding how scent trails work when they are not in the air. If the pheromones are hydrophobic, as has been reported in other snakes, there would be ground trails or trails in the object the snake touches underwater, so male anacondas should be able to follow scent trails underwater just as well as a terrestrial animal would follow a scent trail on land. This is likely what happened with the snake reported in the quote at the beginning of the chapter.

The Dating Game

It is hard to say what is happening in these breeding aggregations because they do not move much, and everything happens under the mud, so observing them is very difficult. However, I managed to collect enough observations over the years to reconstruct at least a partial picture of what happens. Basically, a female is courted by several males for several weeks (see Figure 8.2). During this time the female mates several times, likely with several males. I am currently doing a DNA analysis to see if she mates with several males, but I will be extremely surprised if she did not.

Probably what happens it that a ready-to-breed female lies in the mud or in shallow water, basking frequently and for long periods of time. Once the female becomes attractive, males find her by following some sort of scent trails and start courting her. The female does not become actually attractive, as in putting out pheromones, until she sheds,[77] but a female that is in reproductive condition may be followed by males even before she starts putting out pheromones, as some cursory data suggest (I'll provide more details later). When the female sheds, she puts out the pheromone, several males coil around the cloaca, and the pushing tournament begins. Males coil around the caudal quarter of the female but sometimes cover her almost entirely (see Figure 8.2). While coiling around her body, the males scratch the female with their spurs, moving them very rapidly in a tickling fashion. Males also poke the female with their spurs, inducing her to move and facilitate the position of their cloacae. The male's use of the spurs occurs in bouts, typically 10 to 30 seconds in duration. Often the female moves or twitches in response, allowing the male to continue maneuvering his tail under and around the female's tail.

Each breeding ball was made up of one receptive female and one to several males (up to 13), with an average of 3.83 males per breeding ball (Figure 8.4). The mating season last about 2 months (Figure 8.5), and during this time it is possible to find a mating aggregation that may last as long as 4 or 6 weeks (Figure 8.6). These are several weeks in which the animals do nothing but

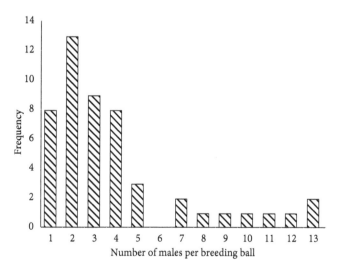

Figure 8.4 Frequency distribution of the composition (number of males) found in the breeding aggregation over the mating season.[18]

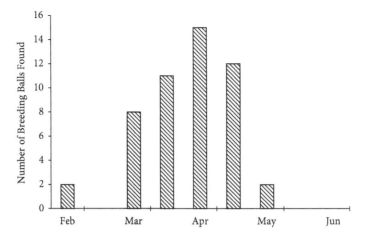

Figure 8.5 Number of anaconda breeding aggregations found, during 2-week intervals, from January to July at the study site in the Venezuelan llanos.[18]

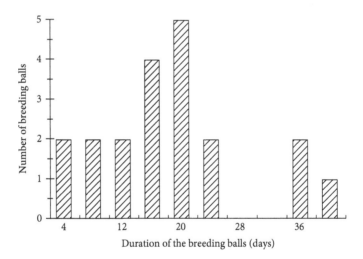

Figure 8.6 Minimum number of days that each breeding ball of anacondas was seen together.[18]

participate in the mating aggregation. There is no feeding during this time, and males seem to be devoted entirely to win the favors of the female. The female may move in or out of the sun, in or out of the water, as she wishes and drags the males along with her. Due to their smaller size, all the males can do is to try to hang on to their position around the female's cloaca.

What If Nobody Comes to the Party?

What if a female anaconda made a big reproductive investment and nobody came to mate with her? Is it possible that she will lose all her eggs and energy? I happened to run into just that situation with one of the largest females of my study site. Marion was a female that I caught the first year and continued capturing again over the years. In 1996, when I was going to do my second year of telemetry, I found Marion again. She was 4.75 meters (15.6 feet) long and weighted 70 kilograms (154.3 lbs). With a condition index (CIA) of 0.973 I expected her to reproduce that year, so I wired her to document her life thoroughly during that year. Marion was in a shallow spot of Caño Guaratarito that was covered with water hyacinth (*Eichornia* spp). Caño Guaratarito was a favorite hunting spot for me because I had found most of the largest anacondas there. It is deeper than the *módulos*; it is a river that runs for part of the year, although in the dry season the ranch managers dam it to retain water and it becomes stagnant. However, even then, the surface is mostly free of water hyacinth.

I have found few males in this river, and most of the females I found were not breeding. They were often in caves at the roots of the trees that pepper the riverbank. Due to its deeper water, Guaratarito is rich in caimans, capybaras, and turtles. For years, I felt this was the place where females go to gather energy, feeding on large prey in order to reproduce or to recover from parturition. Its deeper water and abundance of prey made it a great place to capture, say, capybaras and drag them into the depth of the water, where the snake would be safer from retaliation from the capybara's family. Here is where I witnessed Zuca capture a young capybara and where I saw Olivia kill a very large spectacled caiman.

I was interested in studying Marion for obvious reasons but also because I wanted to see if she would make a breeding ball similar to the others I knew; the low number of males I had seen in the river made me skeptical that males would find her. I searched her location thoroughly, and many times, to see if there were other animals. I was pretty confident that there was no male in that spot of river, and since smaller anacondas always prefer to travel on water covered by vegetation, I was wondering how big a breeding ball she would make. Males would have to take a big risk to get to her since it would involve traveling through water without vegetation in an area with a healthy population of caimans. Flemosito, one of the few males I found in this river, was preyed on more than likely by caimans.[63]

The season went on uneventfully. Every other day, I religiously found Marion and scanned the neighborhood for courting males, but to no avail.

To the best of my knowledge Marion did not mate that year, certainly not the way all other anacondas mated. Was it possible that she mated in a different way? Could a sneaky male have found her, mated in the dead of night, and left, escaping my surveillance? It is possible, but this would have been very different from the other matings I had seen. Why would a male mate and leave if there were no other receptive females in the neighborhood? Plus, Marion was a large female that had lots of eggs. It would be best for him to stay there and secure fertilization of all of Marion's eggs.

I never saw Marion mating, but as the wet season progressed she was acting pregnant: Not only did her condition index indicated to me that she was going to breed, but she was not moving much, staying in the same bend of the river basking frequently, just like pregnant females do. When the time of parturition came, I needed to know if she had mated. Not wanting to try my luck searching for the babies in the wild after she gave birth, I decided to capture her to have her give birth in captivity and to be sure I could capture the babies.

However, Marion was a large and ornery female in a river; that meant we were in for a fight. My assistant at the time was very helpful, but at 50 kilograms (110 lbs) she could only be of so much help. People in the llanos were always scared of anacondas, and it was difficult to find any volunteer to come and help me catch the animals. I went to the tourist station of the ranch where the new tour guide, Rodney, had expressed interest in learning and helping out. Rodney was not a very large guy, but he was fit and interested in helping, which was more than I could get from everybody else. In addition, he spoke English, which would make communication easy if needed, since my assistant was from the United States. The important thing to keep in mind about Rodney is that he was a quiet guy. This is probably a good trait when you are a bird watcher, I guess.

All three of us went in the boat to search for Marion. We brought a 208-liter (55-gallon) metal drum to contain her until we got to the permanent cage where she was going to stay until she gave birth. I found the location of the transmitter. She was underwater. This was not good: She could easily slip out of our reach into deeper water, where we would have very little chance of finding her. This meant it was critical that we got her on the first attempt. Being a multiple-year recapture, she was quick to put up violent fights and try to flee as soon as she noticed my presence. Marion was close to an island in the river in a spot that was shallower than the rest, about 70 or 76 centimeters (2 or 2.5 feet) of water. We located exactly where she was and got off the boat a few meters down the river to get ready. I was going to reach underwater where the transmitter told me she was and try to secure her head, after which my team could step in and help me subdue her.

I reached blindly into the murky water and landed right on top of her. I pinned part of her with my body and my knees, kneeling on the ground on top of her. I tried to locate her neck or head, feeling blindly for a part of her that was thinner or that tapered down. I found a segment of snake that was not very thick, but I wasn't sure if it was close to the neck or close to the tail. If it was the tail, it was a risky situation because she could slip and get out of my grasp. If it was the neck, then I was sitting on the middle of her and I had a better chance of hanging on to her—but again, I was very close to a very toothy mouth of a very feisty, very large snake.

My doubt lasted only until I felt the painful bite of her grabbing my shoulder and upper arm, which were just below the water. "I'm close to the neck!" I thought. Not wanting to relinquish the secure hold I had of her body, I didn't rush to grab the head—I knew where it was, holding my shoulder! However, before I could get ahold of her, she let go of me and continued to thrash around in front of me, trying to swim away. Knowing where her head was, I tried to tell my helpers to go behind me, where they could help me pin her down until I secured the head. "*Detras! Detras! Detrás de mi!*" I yelled, ordering them to go behind me since I knew the head was in front of me. I didn't think in the heat of the moment to repeat my command in English, knowing that my assistant understood only survival Spanish.

Rodney obeyed and went behind me, but my assistant didn't grasp the whole meaning of my words. Seeing me yelling in distress, she only read the emotional language and understood I was about to lose the snake, when I actually was yelling to keep them out of harm's way. She reached in front of me, trying to get the snake that she thought I was about to lose, only to find Marion's open jaws. Marion tagged her on the wrist. I saw her pulling back in pain, with Marion's gigantic head attached to her wrist. I quickly climbed up Marion's neck. Marion let go of her and turned around, biting me on the right hand. I was finally certain where the head was! I reached with my left hand, my dominant one, and seized her neck while maneuvering to free myself from her bite (as described in Chapter 2). I pulled her head out of the water and worked on securing the evil loop. My assistant grabbed the tail and Rodney handled the middle while I held the front part.

I was busy controlling the evil loop with one hand with the other one throbbing with pain. My assistant also was busy, and she and I made sure to keep our bloody hands out of the water. Piranhas are not the killers they are made out to be on TV, but bleeding in the water is asking for it. They won't strip your skeleton of flesh in seconds like in the movies, but they can take a big chunk of flesh with a bite. When we finally got to the boat, I counted to three, indicating to my crew to pull the snake onto the boat. Then we could put it in the drum later. To my surprise, when I tried to lift Marion out of the water, nothing happened; she didn't move. Then I saw that next to me, Rodney, who had not made any sound, was wrapped by Marion's coils from his waist up, around his abdomen and chest all the way to his shoulders. His arms were included in the coils of the snake and thus were of no use. I couldn't see much of Rodney; most of him was encircled by Marion's thick body. All I saw were his eyes, which were open like fried eggs, silently screaming for help as the water closed over him. I pulled up Marion's body to raise Rodney, who resurfaced, taking a big breath of air.

There is something about the coils of a big snake. You can't break into them once she has made them. The way she locks up the coils with her body makes it very difficult to loosen them. It is a bit easier to uncoil them from the tail, working your way up toward the head. Now that we knew Rodney was in trouble, we wasted no time. My assistant started working her way up Marion's tail and loosened Rodney from Marion's embrace. We eventually uncoiled a very scared Rodney and managed to put Marion on the boat and then in the drum, securing a lid over it. To this day I can't understand why Rodney didn't ask for help, or in any way let us know he was having problems. I guess he is a very quiet guy indeed!

I had Marion in a cage for about a month and a half until she gave birth. She had 32 stillborn babies (18 girls and 14 boys) and 16 infertile eggs. Adding stillborn plus infertile eggs, we have a relatively normal clutch for her size: 48. Even though I cannot be 100% sure about her mating, this result suggests that she did not mate. The unusually large number of infertile eggs (one-third) suggest that she did not have a lot of sperm available for fertilizing her eggs. Even females who mated with a single male do not have that many unfertile eggs.

The most likely scenario is that the Marion had stored sperm from a previous mating. Sperm can be stored on occasions for years before it is used. Captive snakes that have not been mated for many years do on occasion reproduce using stored sperm. More recently, DNA analysis has allowed the identification of another way of reproduction that does not involve the participation of a male: facultative parthenogenesis. Females that have not been exposed to a male in many years, or in all their life, may still produce a clutch!

It makes evolutionary sense for a long-lived female to have a plan B for reproduction, whether it be parthenogenesis or sperm storage. The energy used to make the eggs is quite substantial since they are very large eggs (150 grams [5.3 ounces]), there are very many, and they are composed mostly of fat and protein. Losing all that energy would be a great waste for a female that fails to find a suitable partner. So, evolution has endowed the animals with an alternative strategy for "rainy days." Many people find it difficult to get their head around the idea of females reproducing without having sex. This may be partially due to our human bias, which places a very high priority on sex. Also in Western culture, patriarchal relationships are the dominant ones in human societies. So, most people are surprised to see what a small role males have in the cycle of life. Apparently, the main contribution males make to the clutch, in evolutionary terms, is to help with mixing genes and adding genetic variability to the offspring to help with parasites and diseases and protect against random events.[93] In reality, it is quite a waste of effort for a female. She makes such a big energetic investment in each gamete, then develops their babies for several months inside her. Conceding half of the genes for the next generation to the male is a very high price to pay. In the short term, females are much better off, when they have the cellular mechanisms to do it, to make babies by themselves and maximize the number of copies of their genes they leave to the next generation.

In snakes, females are the heterochromosomic sex (the one that has two different sex chromosomes). In mammals males are the ones that have X and Y chromosomes, while females are homochromosomic (XX). The sex of the mammalian offspring is determined by chromosomes carried by the sperm that fertilizes the egg. The eggs always carry an X chromosome. In snakes the situation is more complex. Females are the heterochromosomic sex. Females are ZW, while males are ZZ, so the sex is determined by which chromosome is in the eggs, not the sperm. Basal snakes like anacondas may engage in reproduction without males on occasions. Females, in particular females that have not been exposed to a male in many years, may produce a clutch of babies by themselves. One of the paths by which this occurs results in offspring who are all males.[94] This makes evolutionary sense: If the female has not been found by a male in many years for mating, a safe assumption is that there are not many around. There might be, like her, other bachelorette females nearby. Producing males is the best chance for her to produce grand-children (the ultimate goal of evolution).

Another path for parthenogenesis results in offspring who are all females. This alternative system produces females of the type WW.[95] What this means is that a parthenogenetic snake can bear either males or females. The evolutionary underpinnings of this are not clear. It may be the female trying to have her cake and eat it too. Since it is always beneficial for females to get rid of males, parthenogenesis is always a path they may take. Parthenogenesis is thus a recurrent theme in evolutionary lineages, but often sexual reproduction returns due to the benefits of increased genetic diversity.

At any rate, the case of Marion cannot be one of these. Due to the large number of unfertile eggs and the presence of individuals of both sexes in the clutch, my first suspicion is long-term sperm storage. However, I collected tissue samples from the babies and am in the process of doing the DNA analysis needed to confirm this.

Competition Between the Males

Mating systems in snakes are rather complex. The following are some of the features found in other snakes:

1. Male snakes have been reported to fight viciously, inflicting serious injuries on one another with their teeth and spurs.
2. Some other species engage in ritualized fights where males wrestle with one another, trying to pin down their rival and demonstrate superior strength, but without biting each other or doing each other physical harm. This strategy is common among pit vipers, which have very powerful weaponry.
3. In another kind of competition, males wrestle with their tails, pushing other males out of the way.
4. In some species males gather around the female and mate with her without apparently paying attention to the other males in the area. Allegedly, when several males find a female, if two of them fight over her, a third male may take the opportunity to mate with the female without wasting time or energy in the fight, so males ignore each other and just focus on trying to mate.

The first three strategies involve physical competition and lead to development of large size in males. The fourth one does not involve physical competition but may involve sperm competition. In fact, size alone, while important for preventing other males from mating, is a poor predictor of male reproductive success when the female mates with several males. When several males mate in a breeding season, sperm competition is at work. Natural selection would favor those males that have larger testes or that otherwise develop specialized strategies for sperm competition. Of course, sperm competition may also be present in species that have physical competition, so long as the female mates with several males.

When a female mates with several males, sperm of the males ejaculated in the female's reproductive tract continue to compete, using a variety of methods that range from simple to very sophisticated. Some species have ways to remove the ejaculate of former males in the female's reproductive tract. Other strategies include deploying massive amounts of sperm in the female's reproductive tract. Other methods involve the use of chemicals that produce a net where the sperm of rival males gets stuck, or sophisticated "assassin" sperm that target the sperm of other males for destruction.[96] Ecologically, sperm competition is very important because it removes, or at least lessens, the advantage that some males may obtain by mating more often or sooner. Predictions of reproductive output based on mating often do not hold true when sperm competition is at work.[93]

As long as I have worked with anacondas, I have never seen any actual or ritualized combat, as described in the literature for some other snake species.[97-99] Anacondas do not fight directly with one another, nor do they engage in ritualized combat. This was of no surprise to me as the strong sexual size dimorphism suggested that there should not be any physical competition among the males. Rather, sperm competition must be important. Given that the presence of sperm plugs has been documented, there cannot be any doubt that it happens. However, to say that size doesn't matter would be a mistake. When a new male joins a ball, he coils around the female and starts pushing his way toward her cloaca among the group of other males. Only the male that is wrapped around the female's cloaca has any chance of mating at any given time.

Once a male mates with the female, he leaves a sperm plug that is composed of congealed protein that becomes solid, blocking the cloaca of the female. This may serve several purposes. One of them may be to prevent the sperm from backing up from the proper place when the male pulls out his hemipenes (Figure 8.7). It has been argued that the sperm plug may contain "turn-off" pheromones that deter other males from courting and mating with a female that has already mated. However, the evidence for this is very weak, even in the garter snakes where it was described. Another possible function of the sperm plug is to prevent other males from mating if the mating male is dislodged from his privileged position. Regardless of the modus operandi, so far the data indicate that sperm plugs in snakes are "surprisingly ineffective" at preventing other males from mating.[100]

In anacondas, there is no evidence of "turn-off" pheromones, since I have found many mated females, with sperm plugs intact, in the midst of a fully active breeding ball with males intent on courting and mating with the female. In some occasions, I was able to see mating and the sperm plug left behind. This sperm plug does not last more than 48 hours, after which it comes off, apparently by itself. A female that I held in captivity with all 11 males that were courting her dropped five sperm plugs in a period of 11 days, suggesting that she mated at least that many times. In fact, the very existence of sperm plugs, which are an important element of sperm competition, is evidence that several males mate with a female and that there is strong male–male competition beyond the physical struggle for mating.

Due to the metabolic constrains of ectothermy and the presence of many other males, it is unlikely that a male can retain the optimal breeding position throughout the breeding period of the female, regardless of his size. A possible strategy in the mating of male anacondas is that they alternate physical competition and sperm competition by leaving a sperm plug when they can no longer maintain the prime position. After recovery, males might re-enter the struggle. This scenario would account for the duration of the breeding balls.

Thus, males that mate more often, and prevent other males from mating by maintaining the breeding position for longer and/or by using sperm plugs, would have a higher probability of siring offspring. A male would be best served to mate, leave a sperm plug (for whatever it might be good for), and stay there coiled up to the female's cloaca, preventing other males from mating. Holding his position at the female's vent would prevent other males from mating and would give him another chance to mate again after he has replenished his sperm storage. Given the duration of the breeding ball, it would definitely pay off for the male to try to mate again. So, slow-motion wrestling may occur, with males pushing one another away from the vicinity of the female's cloaca.

Wild male anacondas that mated in captive trials did not leave the breeding ball after successfully mating; instead, they stayed coiled around the females, apparently preventing other males from mating. Clearly, the more times a male mates, the more sperm he would inject and the higher the odds of his sperm fertilizing the eggs. At a very simple level, it is not that different than buying tickets for a raffle. The more tickets you buy, the better the odds you will win. I believe that males may benefit by staying with the female they have found because even at the height of the dry season, females seem to be too dispersed and unpredictable for the male to have a good chance in locating other females in the same season (see later for more details). It makes sense for the males to spend all the time they can courting and trying to mate with the female they have found, since the odds of finding another female in the short breeding season are slim. Figure 8.8 shows the length of the breeding season. Compared with the time the breeding ball lasts, it is clear that a male that tries to court several females has a very small chance of success.

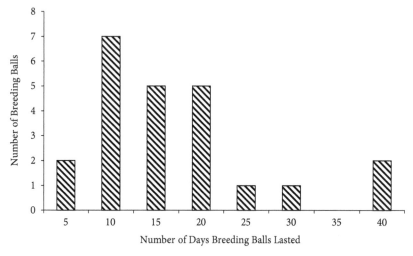

Figure 8.8 Time spent by individual males with a particular female. The time spent was determined by following radiotagged animals or by recatching them at different times. These are observations of males that were already in a breeding aggregation, so the actual time spent in a mating aggregation must be higher than reported.[18]

Seeing the World at Anaconda Speed

One of the problems that hinder the study of anaconda mating systems or possible male competition is that there is nothing for the human eye to see of this competition. Despite long hours of observation of breeding balls, there is not much more to see than some animals shifting or moving in an apparently unpredictable manner. Fortunately, because male anacondas are so intent on mating, they are not seemingly bothered by the presence of people around the breeding ball. I had the opportunity to see how it works by good fortune. A film crew from the BBC came to my study site and wanted to film a time-lapse video of the anacondas since the real-time image looks very much like a still photograph and is not very exciting. I helped them get the footage and asked for a copy for my own amusement. When I got the footage I was astonished to see the action that was going on in the breeding ball. The speeded-up footage shows the bodies of the males moving up and down as they breathed in and breathed out. This movement is not apparent when watching speeded-up footage of lone males courting a female filmed in captivity. So, what one sees in the time-lapse video is actually the males panting, and this panting seems to be due to the presence of other males. They are panting at anaconda speed that is undetectable to our eyes. That the males pant is a very strong indication that there is physical competition among the males.

Brutus or Popeye?

The presence of physical competition among the males presented a problem for me. The strongly female-biased sexual size dimorphism told me that physical competition between the males was not important. On other hand, seeing the males pushing each other in the mating ball suggested that there was indeed an advantage to large size in males. The way to solve this riddle is to look at the paternity of the babies and figure out who the sire is (are?). However, I had all kinds of fun bringing DNA samples from Venezuela to the United States, where I had access to laboratories and equipment. I started trying to obtain the permits in the 1990s. This, I thought, would be an easy task as I had been working for the Venezuelan fish and wildlife service and didn't think it would be difficult for them to write me a permit. However, I met with a colossal level of bureaucracy that would not allow me to do it.

Part of that was backlash from Latin American countries that, after many years of allowing all kinds of extraction and bioprospecting from their lands without any control, now were asserting extremely tight control, to the point of banning all permits and choking international research. Even though I am Venezuelan, moving the samples to the United States was considered a "foreign action," which entailed a mountain of paperwork. They required me to sign a contract committing to give half of my profit from this "bioprospecting" to the Venezuelan government. This was easy to do, because I wasn't looking at making any money, and half of zero is zero. But they required me to pay a $9,000-plus "bail" to ensure I would comply with the contract! Not only had I nowhere near that kind of money, but also the notion that they would ever return the money, given the amount of bureaucracy, was downright ludicrous. Their stupidity went so far as searching my house to document that I had the samples that I claimed to have and that I had collected while working for them! I continued trying to get them to listen to reason, and as the years passed, the laws changed. The new laws required documents that need to be acquired before collecting the data, but since those requirements were not in place when I collected the data, I didn't have them. So, I was considered in violation of the "existing laws" even though they hadn't existed when I did the work.

As I struggled with the matter, I commiserated with my fellow scientist in the Universidad Central de Venezuela, who understood well the idiocy of the government regarding research. One day I was sharing my grief with a former professor, Jesús Ramos, who told me: "Tocayo ['namesake'], don't be such a fool. Smuggle those samples to the United States; nobody will say anything." This was back in the late 1990s, when air regulations and inspection were not as strict in the United States. It was well-meaning advice from someone who both knows the bureaucracy of the government and is familiar with the need to do scientific research. Yet, I decided not to take his advice and continued instead trying to cross all the T's and dot the ever-increasing number of I's.

As the years passed, a new government came to the country and things changed so much in Venezuela that at some point I got wind that the new administration had eliminated a lot of the ridiculous bureaucracy. The new government had put real professionals in positions of power instead of career politicians and cronies, as had been the case in the old days. I went to the administration of biodiversity hoping to find a professional who would understand the scientific merit of my work and would be responsive. To my surprise, when I went to the main office, I saw none other than Jesús Ramos himself, now retired from the university, who had taken a position managing the office of biodiversity. My heart skipped a beat, and as soon as we were seated, I told him, "Tocayo, remember that time you told me to smuggle the samples to the US?"

He nodded.

"Well, I didn't do it, but now that you hold this office, you have to let me get these permits."

Jesús Ramos rolled back on his chair, letting out his characteristic raucous laugh, and consented to grandfather my request since it predated all the problematic laws. He arranged and expedited a permit.

I have now samples in the United States, after more than a decade of struggle, and I am working on the DNA analysis. You can see that I play a very tiny violin when I hear my fellow US researchers complain because getting a permit in the United States may take *several weeks*!

However, in the absence of DNA results, I had to come up with another way to answer the question. In the process of capturing the breeding ball, it was hard to keep track of who was mating, who was in the mating position, or who was simply in a suboptimal position (i.e., far from the female's vent) (Figure 8.9). However, I obtained enough information that I could test statistically what was happening. The largest male was in the mating position significantly more often than the other individuals. Although being in the mating position does not mean mating, and mating does not mean siring, being in the mating position is a requisite for the other two. If a male is very good at sperm competition and very suave in courting the female but is bullied out of the mating position by the other larger males, he still would not be able to court or mate, so his skills will not provide him with any benefit in siring offspring. The presence of the largest males in the mating position does not prove that size give mating advantage but suggests that it might. It also suggests that males may need to be at least a certain size to succeed at siring offspring.

Another strong indication that there is physical competition among the males is the relationship of male size and female size in a breeding ball. Figure 8.10 shows the average size of males in a breeding ball in relation to the female "hosting" it. We see a very clear trend of larger females being courted by larger males. This is called assortative mating, when mating is not random.

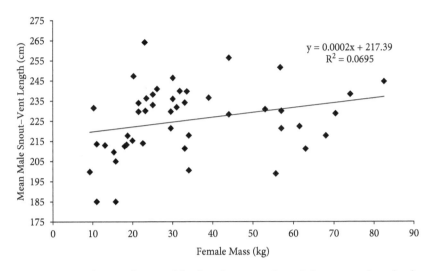

Figure 8.10 Relation of mass of the female anaconda and the average length of the males in the breeding ball ($r = 0.38$; $p = 0.009$; $n = 45$).[18]
Reproduced with permission. Eagle Mountain Publishing, LC.

Assortative mating can occur because, first, males compete with each other and displace the smaller males who cannot hold their own. Second, larger males could also be selecting larger females to maximize their breeding effort, as larger females have much larger litters. If smaller males don't have a chance with the larger females, it would be in their best interest to court smaller ones. Whether it is due to exclusion of smaller males or an ontogenetic switch in mating strategy, the correlation between the size of males and females can be interpreted as further evidence of physical competition among the males. Not only are larger females courted by larger males (see Figure 8.10), but they are also courted by more of them (Figure 8.11). Thus, physical competition over females is definitely present.

Habitat selection is an alternative scenario to explain this assortative mating. It is possible that smaller males avoid the exposed places where larger females hang out and bask, reducing their risk of being preyed on. The use of places with more vegetation or deeper water by smaller animals would produce the apparent association of larger males with larger females. In my observations it was apparent that smaller breeding females were often found covered by mud or aquatic vegetation while larger ones were more often found in shallower water or on dry land, but this trend was not universal. However, it would still mean a larger benefit for larger males, since

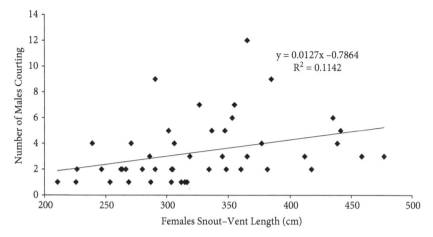

Figure 8.11 Relationship between number of male anacondas in a breeding ball and the length of the female in it ($r = 0.48$; $p = 0.001$; $n = 46$).[18]

larger females produce larger litters.[15] Thus, selection pressure for large size in males still is supported.

There are still other reasons that would increase advantage for large size in males. Larger males have larger testes and consequently more sperm and thus an advantage in sperm competition. Because there is selection for large size in females, the female may actively select larger males that provide genes for large size for her daughters—"sexy daughters."[101] Homologous morphological traits in males and females are expected to show high genetic correlation.[102] This is the reason that men have nipples: Something that is very advantageous for a sex (female's breast) ends up in the other sex even if it is not needed as a consequence of the basic blueprint to make a baby. So, if selection for large size is so strong in females (see Chapter 6), we would expect that males would be affected by that selection for large size. All of these are selection pressures that would lead males to be larger, yet males are substantially smaller.[18,19]

The Boy's Perspective

Physical competition indicates that there must be selection for large size in males. In other words, larger males should be able to produce more offspring. As larger males have more and larger sons, this would lead to the evolution of larger size in males. Yet anacondas have among the smallest males compared to the size of females. The truth about physical contests is that large size is always beneficial but not always decisive. Many times factors such as tenacity and determination can be decisive in the outcome of a match. This is only considering the male part of the story. However, the more we learn about mating systems, the more we learn that female choice is the determinant in most if not all outcomes. There was a time when all science was made by men and when we believed that females had no say in reproductive outcomes or reproductive dynamics. In this view the males "duked it out" among themselves and decided who would mate with the female. The female then would passively accept whoever won the fight. As we have learned more about mating systems, and since women started becoming more active in the scientific disciplines, other views and other hypotheses have been tested. Now we understand that female choice is ubiquitous and very prevalent in the mating systems across taxa.[103]

In one instance, I noticed copulation occurring (i.e., hemipenis inside of the female's cloaca; see Figure 8.7). This copulation continued for 100 minutes. I could see the movement of the male back in the regions where the testes are likely associated to the peristalsis movements in the seminiferous canals associated with ejaculation. As night was falling, I needed to capture the animals to collect body measurements and I tried to remove the male from the female. The male responded by tightening its coils around the female, likely the response intended to rebuke another intruding male. I tried harder and overcame the resistance of the male, either by my superior strength or by my grip, which indicated to the male that I was not a competitor but a potential predator. The male started to uncoil, but at that time the female (Andrea, 307 centimeters [10 feet] snout–ventral length [SVL], 351 cm [11.7 feet] total length [TL], and 23 kilograms [50.7 lbs]) coiled her large tail around the male, preventing me from removing him. Up to this point, the animals, certainly the female, did not seem to be aware of my presence due to their heads and eyes being under the water and mud and the fact that I avoided touching the female to prevent her from fleeing.

The initial resistance of the male to my intruding hand is a strong evidence of physical competition between the males, as he clearly was fighting back. Andrea's holding the male in the mating position could be explained by simply arguing that the female did not want to be interrupted in the act (who does?). But it also suggests that she was not neutral regarding the individual she was mating with. If larger size and strength were all that mattered, the female would benefit from having babies from the larger male, so being interrupted and mating with a larger male would be welcomed by the female, as this means her babies would have better (larger size) genes. Yet, that was not the female's position. Once Andrea had chosen her mate, she did not want a newcomer to remove him even if it was a stronger one who could dislodge the mating male from the her cloaca. Whatever criterion had led Andrea to choose this male, she was sticking to it. In other words, what this observation suggests is that it's not only the males who decide the mating; perhaps it's not their decision at all.

Hey, It's Me, It's Me: Talking Straight in the Breeding Ball

Summarizing the anaconda mating system, we have a strongly male-biased operational sex ratio of 1 female for every 3.83 males. This is not the regular sex ratio that we discussed in Chapter 3; it refers not to the total number of males and females in the population but rather to the ratio of males and females that are actually mating in a given year. Not all adult individuals are willing and ready to mate in every year. As we discussed in Chapter 6, females incur a very large reproductive expense in every reproductive event and they need at least a year to recover from it, maybe more, so the proportion of males per female that is breeding at a given time is not necessarily the same as that found in the demographic study. To the male-biased sex ratio that we found in Chapter 3 we add an even stronger bias because about half, or more, of the females are not available every year.

Because of this biased operational sex ratio, we expect the competition among males to be intense; there are few females for all those males, so females become a scarce resource for which males compete. We have learned that females may mate with several males but males would benefit most from hogging all the mating possibilities and excluding other males so they will sire more of the female's babies. It is not clear how this competition happens. There is certainly some sperm competition, but there is also a great deal of physical competition. This latter is expected to lead to the evolution of larger size in males, but this doesn't fit with the very small size of male anacondas compared to females.

One day I tried to put into practice the notion of "wearing the snake's shoes," trying to figure out what was happening in the breeding ball. I wondered if it was any easier for the anacondas themselves to know what was going on in the ball. I tried to see what a male anaconda sees in the breeding ball. If I were a male anaconda and were in the breeding ball, with my body wrapped around the female and other animals, would it be easier for me to know what was happening? Everybody is covered with mud and aquatic vegetation. How do they know who is who? It occurred to me that if I were there, I'd have to rely on tactile cues to find out how to court the female and not another male. Given, I don't have the formidable sense of smell that male anacondas have, but with all the males wrapping around the female and the animals rubbing against one another, are chemical cues meaningful anymore? In fact, when we look at a breeding aggregation, it's clear that males' heads (where their eyes and chemoreceptors are located) are in one direction while the tail is attempting to mate completely out of sight and away from the reach of chemical senses (Figure 8.12).

This was an exciting realization. I knew that I could assess the sex of an anaconda at a glance, provided that it was an adult, based on how much thicker females are. The sexual size dimorphism is so dramatic that I only need to look at the snake's girth: If it's broader than my forearm, I know it's a female and if it's narrower it's likely a male, provided that it is an adult. If I can use the size of the animal to identify the sex, can't the anacondas use it as well? I set out to look for evidence that anacondas use tactile cues to distinguish the males from the female in the breeding ball. On several occasions I observed males coiling around the female's neck, apparently courting the "wrong end" of the female (Figure 8.13). The confusion of the males wrapping around the female's neck is consistent with the hypothesis that they use tactile cues to find the female's cloaca in the breeding aggregation. Clearly, there is no reason for the female to have any chemical signal leading males to court her neck! The neck is the one other part of the snake body that becomes narrower. If the males used the girth of the snake body as a clue, we would find this kind of confusion, and we are finding it.

This was encouraging but not conclusive. I continued looking for evidence of males using girth for sex identification. On at least four occasions I saw some males coiling and "courting" around the tail of a large male, apparently confusing him with the female due to his large size. In one instance a very small female (275 centimeters [9 feet] TL, 11 kilograms [24.3 lbs]) was being courted by four males, two of which were relatively large specimens (277 centimeters TL, 8.75 kilograms, and 280 centimeters TL, 8 kilograms), which were in turn being courted and coiled by other males in the breeding ball. Perhaps the most dramatic example of this occurred with Ashley's breeding ball. Ashley was a very large female that was with 11 suitors. I put her in a naturalistic enclosure with a pool and dry land for basking so I could better observe the mating dynamic. At some point, she was in the water with her host of suitors and decided to come out of the water to bask. As she moved, males that were wrapped around her cloaca were dislodged from their position. Figure 8.14 shows two pictures of the enclosure. In both pictures, it is possible to see a group of males wrapped around Ashley's tail (marked with the letter A) while other males are wrapped around the tail of another very large male (B) that apparently was unable to follow Ashley in her movement due to the other males that had mistaken him for a female.[19]

This finding suggests that males may have to pay a price for being too large. Clearly, if a male is being mistaken for a female in the breeding ball, he will have to spend time and energy in freeing himself from the males that are courting him, which handicaps his possibility of courting the female himself. So, male anacondas could be under pressure for large size because larger males can push away smaller ones from the mating position—but there may be a size that they do not want to pass or else they would lose mating opportunities if they are mistaken for females. So, male size may be under stabilizing selection for an optimal size that is big enough for him to hold his own in the mating aggregation but not too large to be confused with a female.

I then set out to look in my data for evidence of this stabilizing selection. Stabilizing selection, when present, produces clear and recognizable effects on the population. It will produce stereotypical results on the trait in question. In the case of size, stabilizing selection will produce individuals of a very specific size with very little variance. I looked back at my data and was fascinated to find exactly that. Figure 8.15 shows the size distribution of all animals found in 50 breeding aggregations. Here it is possible to see something that I knew, but I was still surprised to find that it was so obvious. The first observation is that it is possible to determine the sex of the animals with high accuracy just by their mass. Males grow pretty much up to the size of females when they start reproducing, but then they seem to stop while females continue to grow. The other aspect is that most males fall between 4 and 8 kilograms (8.8 and 17.6 lbs) in mass, while females have a lot larger variance than males, falling between 12 and 80 kilograms (26.5 and 176.4 lbs). This extremely large difference supports the idea that males are under stabilizing selection for an optimal size. These two pieces of evidence support strongly the idea that anacondas use size as cues for sex discrimination in the breeding ball and explains the extreme sexual size dimorphism found in

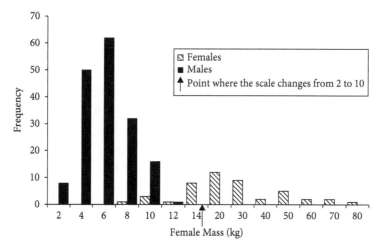

Figure 8.15 Size distribution of animals found in breeding balls. Notice the change in the scale after 14 kilograms (30.1 lbs).[19]

Reproduced with permission Elsevier.

anacondas. Females are under directional selection for large size, as seen in Chapter 6, while males are under stabilizing selection for a convenient size. So, the optimal size for a male is one that allows him to outcompete rivals and hang on to the female's cloaca but that does not mislead other animals into mistaking him for a female.

How Did This Sexual Size Dimorphism Come About?

Female-biased sexual size dimorphism is the most common scenario among snakes, including basal Macrostomata,[69] and may be the ancestral condition among the group.[15,90] A possible explanation for this extreme sexual size dimorphism is that the reproductive investment of the males may be relatively low and they do not forfeit much growth by breeding early.[40] Thus, males would benefit from maturing early in life as this would give them some chance of early breeding.[104,105] The idea that males divert their resources into reproduction as opposed to growth early in life explains the smaller size of males. Females, on the other hand, need to make sure they have at least some size in order to be able to physically develop the offspring that is large enough to survive. If the baby anacondas have to be at least so large, females have to be big enough to produce and develop a baby of that size in their bodies.

Once we have this scenario (males breeding sooner and being thus smaller), then it is easy for it to be used as an evolutionary clue for female identification in the breeding ball.

The possibility of cannibalism is another important element that might influence sexual size dimorphism, since mating females occasionally eat adult males.[34,35] Presumably, larger males would be better able to prevent themselves from being eaten by females (Figure 8.3). However, the size difference between males and females is such that an increase in male size might not be sufficient to save him; female anacondas are known to eat relatively large prey.[15,24] So, being too large to fight off a hungry female is perhaps out of the question. A male would be better served by being smaller, which means he is not such a good meal. Female anacondas drop smaller meals from their diet (see Chapter 4). A small male that isn't worth the effort for the female may be a strategy leading to small size. Male size also might be constrained for the other reasons such as mobility. With a small body, it's easier to move around looking for females.

In some species dimorphism occurs at birth, where typically males are sometimes born slightly smaller than females.[105,106] This does not seem to be the case in anacondas, where birth size does not differ between the sexes in terms of length or mass.[15] An explanation for the proximate mechanism of the strong dimorphism and small variance of male size is that males and females have different food preferences. The diet of small anacondas primarily consists of birds. While males maintain this diet throughout life, females feed mostly on birds only until they reach reproductive size, approximately 3-plus meters (9.8-plus feet) TL. Around the time she reaches sexual maturity, the female's diet switches to mammal and reptile prey (see Chapter 4); this switch in diet might be related to the increase in energy that allows her to grow larger and to keep up with the energetic demands of reproduction.[15] Thus, a possible proximate mechanism by which males maintain the smaller size is by maintaining a lean bird-based diet and a relatively low feeding frequency. This is supported by parallel mark-and-recapture data showing that many adult males do not experience any perceptible growth in as many as 11 years (see Chapter 4). In fact, a captive-born male at the Bronx Zoo, fed regularly with mammalian prey (a female diet), developed an exceptional mass of 40 kilograms (88.2 lbs; William Holmstrom, personal communication), much heavier than any of the wild individuals we studied. Sadly, the origin of this captive animal is unknown, since it came from the pet trade, a donation to the zoo.

So, the sexual size dimorphism is probably created and sustained by differential diet and serves the evolutionary purpose of assisting sex identification in the breeding ball. Females may delay maturation since they make a larger breeding effort. Males start breeding earlier and at a smaller size than females, since females need to gather more energy to start breeding and the fecundity-independent costs of reproduction are too high to produce a small clutch.[40,107] This differential maturation sets the scenario for natural selection to act, and sexual size dimorphism can be selected as a method of sex discrimination. Once the females are larger and thicker, the stage is set for natural selection to target sexual size dimorphism as a mechanism of sex discrimination.

What a Catch! Who Chooses Whom

Whether a sex is choosy or not depends largely on how much it invests in a mating event. In the typical case of mammals, the male produces very small gametes that he transfers to the female during mating, and that may well be the extent of his commitment to that reproductive event. In these cases, males would benefit from mating as many times as they can with as many females as possible without being very choosy. Females, on the other hand, make a much larger gamete, and in case of a viviparous female, she will have a longer-term commitment to that offspring during pregnancy and upbringing until weaning time, as in the case of mammals. Evolutionary theory predicts that these females must select carefully who they mate with: Mating with a low-quality mate would be very costly as she will have to commit a lot of energy, time, and resources to that low-quality offspring. So, these females are expected to be choosy and make sure the male(s) they mate with is (are) good-quality individual(s) that will give good genes for their offspring. When the female produces several offspring, and several males could sire offspring, she still wants all of them to be of high quality.

This trend is far from universal, however. There are plenty of cases in which the males as well as the females are choosy, and there are other species with complete sex-role reversal where the males are choosy and take care of the offspring, committing large amounts of energy and time to every reproductive event, while the females are philanderers that mate with several males, committing relatively fewer resources to every reproductive event.

When there is sex-role reversal often females fight over the males, grow to larger sizes, and develop weaponry for sexual combats, just like males of other species do. However, the general trend is that choosy individuals seek to mate with partners of high quality even if it compromises the quantity of other mating events they can have. In the cases where the males invest a fair amount of resources in a breeding event, they are expected to be choosy as well. Males anacondas spend quite a bit of effort courting a single female, so natural selection would lead them to be choosy.

Looking at the details of the breeding aggregations, we find that larger females are courted by more males (Figure 8.11). Females of higher quality (larger) are more desirable to males. If males mated with any female they encountered, there would be a random distribution in the number of male and female size. Because larger females produce larger clutches of larger offspring, we can see this as evidence of males choosing a high-quality mate. Males are choosing quality, even if this means stiffer competition.

How Expensive a Ring Should He Buy?

Are the males making a large breeding investment per individual female? The sperm plug that was removed from the female's vent relatively intact represents 0.1% of the male's mass. This is fairly high if we consider that the sperm plug is not all the sperm that the male ejaculated and also that he has only a short time to replenish his sperm reserves within the mating season. It is not all the investment he makes, since he expends a lot of energy searching for females and risks predation in the process, including being preyed on by the female herself![34,35] In the llanos, the breeding period for anacondas is restricted to the driest part of the dry season (approximately 2 months), but the great majority of the matings occur within 1 month, the period when males can most easily locate receptive females. So, courting a female for several weeks represents a big investment because there is not much time left in the breeding season.

I have found no evidence that males philander among different females. The time a given male spends with a female seems very large in comparison with the time the breeding season lasts, with males spending more than 3 weeks with a given female (see Figure 8.8). Considering how widespread breeding aggregations are in the landscape and how unpredictable their location is, it is unlikely that a male would have time to find and court more than one female in a single breeding season. So, anacondas show a polyandrous mating system, which is in apparent contradiction to the mating system reported to be common among snakes.

It is possible that the evolutionary reason for this system is the result of five factors:

1. Larger females produce more pheromones and lure more males toward them, in this way manipulating the males' behavior.
2. Larger females are easier to find because they place themselves out in the open, and because of their sheer size, which means they are encountered by more males.[108]
3. Males look for females of a particular size in the areas where they occur most frequently (larger females in shallower water, smaller females in relatively deeper water).
4. Males prefer particular females based on their private experiences.
5. Qualitative differences in the scent of larger females[109] trigger a mechanism in the male's vomeronasal system or brain that allows him to know the quality of the female.[110]

When we realized that female snakes could mate with several males, the question became more precise: Does the multiple mating mean multiple paternity? Possibly due to male bias among snake biologists, multiple mating in males has often been assumed and seldom documented while multiple mating in females needs to be proven beyond any doubt. Female multiple paternity was accepted only when molecular or genetic evidence showed incontrovertible evidence of that effect. Only by proving multiple paternity through DNA analysis can a female be considered to have mated with multiple males. However, seeing a male courting two female garter snakes in captivity was enough "proof" of polygyny in the species. Conversely, seeing a female courted by dozens of males was not considered evidence of multiple mating in females. I confess to chuckling a little at the irony that it was a hairy-chested Latino who blew the whistle on this bias among snake biologists![90]

Because of the ease with which we consider males polygynous, I was intent in documenting *multiple maternity* using the same standards that are used to prove *multiple paternity*. In other words, I did not want to call anacondas polygynous unless I could document that they actually mate and sire neonates in different females. I did a good deal of tracking males in the breeding season. I employed a quick and dirty way to wire a snake instead of doing the traditional surgery of implanting the transmitter in the body cavity (as reported in Chapter 3). To increase the number of animals I studied in the mating season, I found anacondas, mostly males tracking females, and force-fed them a transmitter (see Figure 8.16). Because they do not feed in the mating season the transmitters last inside them for up to 2 or 3 months, until the end of the mating season, when they resume feeding and pass the transmitter.[111] Despite my efforts to follow radiotagged males in breeding aggregations, I did not document any instances of males joining more than one breeding ball during the same year. Most radiotagged animals that were in one breeding ball left that ball and coiled quietly under the aquatic vegetation without moving after the breeding ball dissolved. Not only did I not document males joining other breeding balls after they had participated unperturbed in one, but also they did not show any trailing behavior, nor did they show the enhanced mobility that characterizes males that are looking for breeding balls. When the results of the paternity analysis are done, I may end up with different results, but so far as I can tell from their behavior in the field, it seems that, unlike females, males mate with only one partner in a given season.

A Story of True Love?

There was a remarkable event when I found Diega. She was eating a young capybara the day the *National Geographic* crew arrived to make the first documentary we did back in 1997 (see Figure 5.14). As I was maneuvering around her telling the film makers what was happening and trying not to disturb her meal, I stepped on something that felt funny. Reaching down, I realized that it was the tail of a male that was one meter away from the feeding female. I took him in for measuring and force-fed a transmitter into him to track his movements. I named him Oops due to his accidental discovery. After I released Oops, Diega was no longer where she was, and I might have interrupted a romance that was waiting to happen, but since he was wired I could follow him through the season. Oops traveled through the swamp for the following month and a half. He traveled just a few meters away from active breeding balls that I was studying but showed no interest in these breeding aggregations. About 1.5 months after his capture I found Oops coiled up with Diega who, by now, was receptive and hosting a breeding ball of her own. If there is a case of true love among snakes, I would say this is it! That year Diega was the largest breeding female I found in that area, so it made sense for Oops to seek her out and not spend time and effort courting suboptimal females.

The first time I found Oops next to Diega, she was not receptive (in other words, she was not hosting a breeding ball). Is it possible that he was there waiting for her to become receptive? Did he know her personally (or snakely, as the case may be)? Or did he just know that there was a large female nearby and he waited for her to become available? As usual, after a lot of research and fieldwork, I ended up with more questions than answers. There is not a boring day in the life of a scientist!

The Girl's Perspective

Female anacondas often mate more than once in a season, and potentially with several males. Females are not expected to engage in multiple mating when a mating event increases the risk of injury or death during or after the mating.[112] The large size of female anacondas, however, lowers the risk of injury while mating. This is consistent with the sedate behavior exhibited by large breeding animals and by the increased exposure of larger females when basking and lying on dry land or in very shallow mud. Multiple mating and insemination has been reported in several species of snakes.[72,100,113,114]

It has been reported in other snakes that the quality of the offspring had a stronger effect on the reproductive success of the female than the quantity.[109] If so, selecting for good mates may be a critical trait for the females. Female choice has been reported as an important issue in the mating system of snakes.[114] Male snakes are apparently unable to forcibly copulate with females due to the elongate shape of the body.[69] It has been argued that a snake's hemipenis is not designed for forcible penetration by thrusting.[115] For instance, in the genus *Epicrates*, copulation cannot be accomplished if the female does not open the cloaca to allow intromission.[98] A similar phenomenon seems to occur in other species such as *Agkistrodon contortix*[116] and *Crotalus atrox*.[97] Females are known to be highly selective in mating aggregations; *Thamnophis marcianus* females reject some males, even after intromission has occurred.[117] While breeding several generations of *Thamnophis melanogaster* in the laboratory, it has been reported that sometimes females accept some males and not others. However, recent evidence suggests that forcible copulation may be possible in garter snakes that mate in very large breeding aggregations at the extreme edge of their distribution range, where females emerging from hibernacula are too cold to resist males.[118] It is uncertain how important this may be among garter snakes, as most do not mate in such spectacular aggregations or in very cold climates.

Regardless of the situation in garter snakes, female choice may be even more important in robust constrictor species where the females are much larger than the males, to the point that they are capable of cannibalizing males.[34] The massive size difference between males and females makes it highly unlikely that males can coerce a female into mating. On many occasions while trying to collect the males of the breeding ball the females was spooked by our proximity (typically if they had been caught before) and slipped through the embrace of all the males. This leads us to think that she was there willingly. Furthermore, our observation of Andrea actively preventing the mating male from being removed suggests that the female choice is an important part in the process.

Long Engagement versus Eloping

One thing about the anaconda mating system that is striking and rather uncommon is the long courtship and mating period. The breeding ball last for several weeks. Why do breeding balls last so long? Can't they mate in a shorter time? A lengthy mating period involves larger exposure to predators and reduces foraging efficiency; for males, it means they forfeit other mating opportunities. There are several possible scenarios that would explain these prolonged mating aggregations.

First, the female may be randomly receptive throughout the period and mate with different males without much discrimination, in which case she is simply encouraging sperm competition among the males. The males that can produce more sperm or sperm that is more competitive will sire the babies.[55,79,119] Lying in shallow mud making a breeding ball that is conspicuous to predators, she is safe due to her larger size, but the smaller males are at risk of predation. Thus, this strategy would reward the big ones that survive predators or the ones that take the chance. This is an example of the handicap principle, which proposes that females select males that handicap their own chances of survival in order to reproduce, which in this case also selects for larger or bolder males.[120] She could also entice courtship among many males that compete in searching skills (finding her first), physical struggle, and perhaps some other way to allow her to select the best males using some criteria unknown to us (perhaps assessing some aspects of the courtship related to the spur movement). What does a female anaconda consider charming?

Another possibility would be that she may not be initially receptive to the males but puts out pheromones and attracts several males that would compete physically for her. She lets the males "duke it out" and becomes receptive for mating only at the end of the period. By doing this she would be selecting the stronger males that have endured the struggle. Some theoretical models predict that males that mate last obtain reproductive benefits as females have a receptacle for sperm storage and the last sperm to go in is the first one to come out to fertilize the eggs. This model may explain the large amount of time the males spend in the breeding aggregation. There might be an advantage for the males to stay after they mate in order to prevent other males from mating later or to mate again if they can—a "last-man-standing wins" kind of scenario. However, due to the previously mentioned cannibalism, there might be a greater risk for the last male that copulates.[34,35] This imposes a dilemma for the male. The longer he stays with the female, the more likely he is to sire her offspring and prevent others from doing so, but also the higher the risk that he might be eaten by the female at the end of the breeding period. This selects for refined abilities of the males to detect the mood of the female, and an abrupt dissolution of the ball as opposed to a gradual one. The field data give ambiguous support for the latter hypothesis. Some field observations suggest that the ball dwindles down gradually, leaving one or two, often smaller, males that still court the female at the end of the period. But there are quite a few breeding balls that do experience sudden dissolution.

Another explanation to consider for the long duration of the breeding balls is the low probability of encountering other females. Females are not clumped together and have a rather unpredictable distribution. Thus, males looking for females must travel relatively long distances, during which they face a high risk of predation. It is possible that it doesn't pay for the males to leave the ball and try to find another female. It is possible that they are better off trying to secure some offspring from one female once they find one. If the chances of the male finding a female are low, the male should improve his chances by sticking with her and not philander around.[121] In fact, in other snakes it has been found that males were more insistent in courting a female in areas with a lower density of animals (and thus a lower encounter rate).[122]

As things stand right now, I feel I have a good understanding of the mating system of the anacondas and only need the results of the DNA analysis to confirm what I have learned. I am pretty certain that the mating system is polyandrous, where males invest a lot of energy per mating and thus try to find the best female they can mate with, while females mate with several males in the same season. I am also pretty confident of the explanation that I have found from the extreme sexual size dimorphism in the species. Females grow as large as they can, which increases their fertility, while males grow up to a convenient size where they can outcompete rivals but are not mistaken for females. All these will, or should, be confirmed with the results of the DNA analysis. Once we know for certain who the most successful male is, we will be able to see if his success is related to persistence in the breeding ball, early arrival, late departure, or some other trait of the courtship. Since I have a few sperm plugs that were obtained from the females' cloaca, we will be able to know if the male that left the sperm plug was more successful or not. We also will be able to know if medium-sized males are more successful in siring offspring than bigger or smaller ones. This would confirm that there is a convenient size for males' reproductive success. As you can see, the fun has just began!

As we are planning to do this project, I can't help but feel some trepidation. It's like building a house of sticks and getting ready to put the last stick on the roof after a long, painstaking effort of building the whole house. Putting the final stick in place can finish up a major, worthwhile undertaking, but it can also crash the whole structure, which would have to be rebuilt from scratch if it turns out that the DNA results do not support all these conclusions. Oh, well, that is the beauty of science! Plus, who wants a house with a sucky foundation anyway?

9

The Origin of the Mystery

Adapting to Life in a Big Dam

At first there was only water. Then a big anaconda came along from the east carrying on her back the seeds of the people, plants, and animals.

Legend of the ancestral anaconda from Tukano people, natives from Colombian Amazon[123]

Un dia Hui'io (la doncella hermana de la Luna convertida en serpiente) salió del agua y desplegando su enorme cuerpo dijo: "Quiero tener una corona" y buscó muchísimos pájaros que cubrieron su cuerpo con plumas, lo que creo a su vez Huasudi, el Arco Iris. Siempre Huasudi saldrá después de una lluvia, ya que la Gran Serpiente arqueara su cuerpo para secar su corona de plumas al sol.

(One day Hui'io, the maiden sister of the moon, turned serpent, got out of the water displaying her gargantuan body and said, "I want to have a crown," and sought very many birds that covered her body with feathers, which created Huasudi, the rainbow. Always Huasudi will come out after the rain, as the Great Serpent will arch her body to dry out her crown of feathers in the sun.)

Makiritare legend[12] (Translated by JAR)

Reflecting on what I have learned over the years, I feel I have obtained a reasonable understanding of the lives of anacondas in my study site. However, since I have done most of my research in a single location, I have to wonder what the life of anacondas is like in other areas where they live. As indicated earlier in the book, I chose the llanos because of its unique dry season and strong seasonality that made the work easier for me. But would these

Figure 5.11 Large anacondas prey on large mammals and reptiles. (a) Capybara (photo Carol Foster); (b) spectacled caiman (Tony Crocetta); (c) side-necked turtle (photo Jesus Rivas); (d) green iguana (photo Victor Musiú Delgado).

Figure 5.13 Capybaras live in social groups. When they are in danger, normally the dominant male gives an alarm call, at which all the youngsters and females dive underwater. On occasion they can attack a predator that has seized one of their group. The largest rodent in the world can swim very well and hide underwater to avoid predation. (Photo Bill Holmstrom)

Figure 5.14 Diega constricting a juvenile capybara (13 kg) in the swamp. Notice the wounds the capybara inflicted on the anaconda during its final struggle. (Photo Carol Foster)

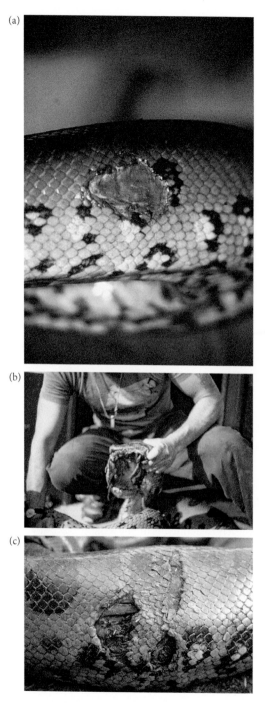

Figure 5.15 Anacondas suffer a variety of injuries when they try to attack a prey. (a) Anaconda with what seems like a capybara bite. (Photo Robert Caputo) (b) Francis after consuming a large side-necked turtle. (Photo María Muñoz) (c) Wounds by, likely, a spectacled caiman. (Photo Bill Holmstrom)

Figure 5.16 Extremely distended skin of a female anaconda consuming a white-tailed deer. (Photo Jesús Rivas)

Figure 5.17 Bottom-dwelling spiny catfish may be easier to catch underwater as they inhabit crevices and corners, but anacondas do not seem to be fit to use them as regular prey. This anaconda tried to eat a very spinous catfish. The spines and shield of the catfish made a hole through the snake's esophagus, muscles, and skin. I removed the catfish through the hole that it had made with very little effort. (Photo María Muñoz)

Figure 5.20 Birds in the drying river. The abundance of birds in the swamps makes it difficult to imagine that anacondas can ever lack food. (Photo Jesús Rivas)

Figure 6.1 Anaconda basking on termite mounds (Judy-Lee) during the wet season. (Photo Robert Caputo)

Figure 6.2 Because anacondas are so charismatic, tourist operations often capitalize on their presence to impress their customers. (Photo Christian Dimitrius)

Figure 6.3 Enclosure where I kept female anacondas during pregnancy. It was a leftover from a failed caiman farm and contained a center pool with aquatic vegetation for them to hide and a dry area for them to thermoregulate and was surrounded by a screen so that the neonates could not escape. (Photo María Muñoz)

Figure 6.7 Ultrasound of anacondas at an obstetrics and gynecology practice. Left to right: Jesús Rivas and Tito Barros. (Photo Maria Muñoz)

Figure 6.9 Babies were released in the field in the place where their mother was caught or in the closest suitable habitat. (Photo Ed George)

Figure 6.10 Female anaconda giving birth. (Photo Carol Foster)

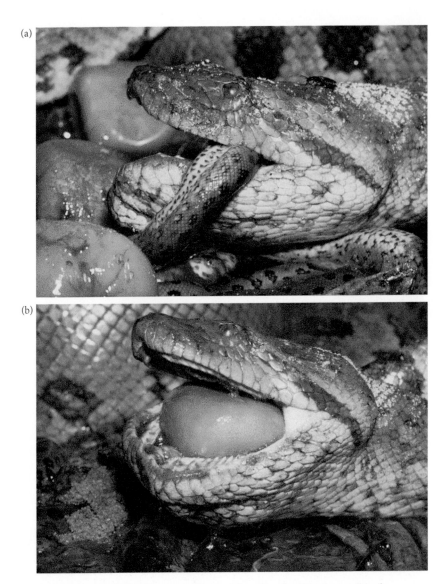

Figure 6.13 Female feeding on stillborn (a) and unfertilized eggs (b) after giving birth. (Photo Carol Foster)

Figure 6.26 Hairball, the newborn capybara I adopted during my time in the field. (Photo Jesús Rivas)

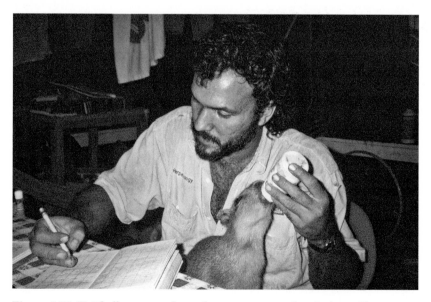

Figure 6.27 Hairball was very dependent on my care when I adopted her. (Photo Jesús Rivas)

Figure 6.28 Hairball wanted Chuka to adopt her, but she didn't feel like it. (Photo Jesús Rivas)

Figure 6.29 Capybaras are very abundant in the llanos, certainly near the roads and houses where they seek shelter from predation by big cats. (Photo Jesús Rivas)

Figure 6.30 Hairball years after I rescued her while she was a little one. Here she is fully grown and belonging to a family of capybaras of her own. (Photo Ed George)

Figure 7.1 Permanent rivers often have larger animals. I found the largest males in my study in rivers of permanent water. This male was 372 centimeters (12.2 feet) total length. Rio Formoso, Cerrado, Brazil. (Photo Sarah Corey-Rivas)

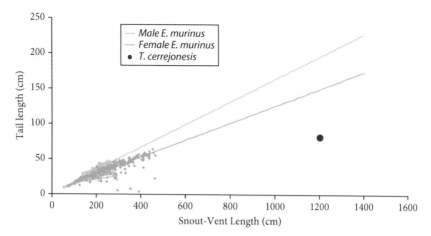

Figure 7.2 Ontogenetic relationship of tail size and SVL in anacondas. The purple dot represents where Titanoboa would fall in this regression. Notice that Titanoboa seems to have had a substantially shorter tail for its size than current-day anacondas.

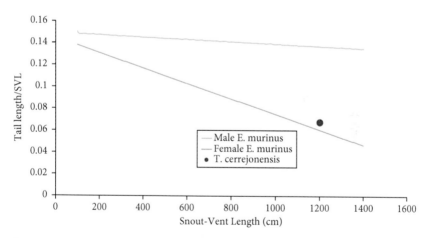

Figure 7.3 Relationship between relative tail length and SVL in anacondas. Within the size range of current-day anacondas it is not obvious to see the shortening of the tails as they grow larger. However, extrapolating it to larger sizes, it is clear that tails become proportionally smaller the larger the snake becomes. This chart shows that Titanoboa's proportions were just on par with those of current-day anacondas.

Figure 7.4 Ashley was approximately the size that a Titanoboa would have been when she started breeding. (Photo Tony Crocetta)

Figure 7.5 Neonatal Titanoboas are estimated to have been the size of these adult anacondas. Left to right: Jesús Rivas and Reneé Owens. (Photo Robert Caputo)

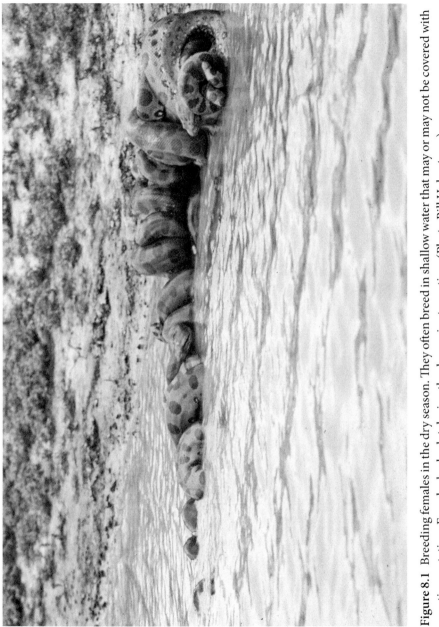

Figure 8.1 Breeding females in the dry season. They often breed in shallow water that may or may not be covered with aquatic vegetation. Females bask a lot during the days prior to mating. (Photo Bill Holmstrom)

Figure 8.2 Aggregations can comprise up to 13 males per female. (a) Benita had 12 males (photo Tony Rattin). (b) Females are often courted by several males usually at the edge of the water or in shallow water with aquatic vegetation. (photo Philippe Bourseiller).

(a)

(b)

Figure 8.3 Female anacondas may, on occasion, make a meal of one of their suitors.[34] This increases the risk of males staying in a breeding ball too long and selects for males that can detect females' moods. (Photo Luciano Candisani)

Figure 8.7 Hemipenes in a male whose courtship was interrupted. (Photo Sarah Corey-Rivas)

Figure 8.9 Catching males in a breeding aggregation often resulted in some of us being bitten since the focus was on capturing them all, and there were many mouths in the same area where we had to collect the males. (Photos Robert Caputo)

Figure 8.12 Mating aggregation of green anaconda. The heads of the males are visible above the water's surface while their tails are trying to mate with the female. In this picture four males have their heads out of the water. (Photo Tony Crocetta)

Figure 8.13 Breeding ball found in the wild with 11 males. Toward the left of the picture is the tail of the female wrapped up by several courting males. However, to the other side of the snake there seem to be an equal number of males courting "the other end," seemingly misled by the thinner girth of the snake in the cephalic end.[19] (Photo Jesús Rivas)

Figure 8.14 Mating aggregation of a large female, Ashley, with 11 suitors. "A" indicates a group of males wrapped around Ashley's tail. "B" indicates a group of males coiled around the tail of a very large male that was courting Ashley.[19] (Photos Jesús Rivas). Reproduced with permission Elsevier.

Figure 8.16 Force feeding transmitter to tracking males was an efficient way to increase the chances of finding breeding females. (Photo Tony Crocetta)

Figure 9.1 Pilcomayo River, which, having been dammed, now covers a broad area, creating what is called Bañado La Estrella in Formosa, Argentina. The paleo-Amazonas would have done something similar on a continent-wide scale. The rising of the Andes dammed the paleo-Amazon. The mega-river would have flooded the entire neighboring area, creating a large swamps. (Photo Jesús Rivas)

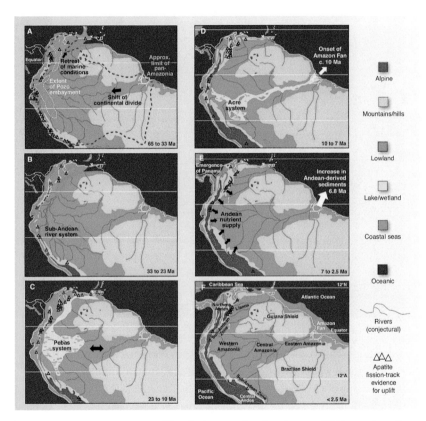

Figure 9.2 Reconstruction of South America's paleo-history.[131] Reproduced with permission. Sciences.

(a)

(b)

Figure 9.3 Cook's tree boa head. Notice the frontal eyes and labial pits, useful for feeling the heat of prey at night. (Photo Jesús Rivas)

Figure 9.4 Yellow anacondas are a bit smaller than green anacondas and are distributed to the north of South America. They are more of a savanna, high-seasonality species. (Photo Nick Lormand).

Figure 9.5 Beni anaconda, *Eunectes beniensis* (a), is similar to green anacondas in the body pattern but lacks the orange coloration in the head (b). (Photos Sarah Corey-Rivas)

Figure 9.7 The anaconda's pattern seems perfectly matched to merge with aquatic vegetation. (Photo Philippe Bourseiller)

(a)

(b)

(c)

Figure 10.1 Anacondas are harvested traditionally through their range, but this likely is not a strong pressure on the population since these are low-impact operations. (Photo Sarah Corey-Rivas)

Figure 10.2 Bringing exotic animals to habitats where they do not belong can cause great conservation problems. The new species has no predators and is able to increase its population without check. The local species do not have any previous experience, or evolutionary adaptations, to deal with the newcomer, and the latter ends up exerting undue predation on local species. This is a Burmese python (*Python molurus*) that has been introduced in Florida. Pythons have established successful populations and have exerted predation pressure on local wildlife.[144] Deer Key, Florida (Photo Ed Metzger III)

Figure 10.3 Anacondas that come near houses and prey on livestock are often killed. (Photo María Muñoz)

Figure 10.4 As humans encroach on wildlife habitats the conflict will increase, and wildlife always is at the losing end. (Photo Jesús Rivas)

Figure 10.5 (a) Thanks to the charisma of anacondas, film crews from all over the world have come to film them, bringing publicity, environmental awareness, and the hope for a truly sustainable use of the species. Left to right: John Thorbjarnarson, Reneé Owens, Jesús Rivas, Richard Foster, and Dennis Town. (Photo Carol Foster) (b) Ed George, Gringo Loco, filming spectacled caimans in the riverbed. (Photo Jesús Rivas).

Figure 10.6 (a) Thanks to the charisma of anacondas, tourists from all over the world have come to meet them up-close and personal. Left to right: Roseanna Faber, Jesús Rivas, Francis Faber. (b) Jen Moore. (c) Front to back Roseanna Faber, Francis Faber, Lisa Bentson, Ana Caudillo, Victor Musiú Delgado. (Photo John Dunbar)

Figure 11.1 Outdoor enclosure where Mi Mujer was being held. Since the cage was older, the wire had holes big enough for Mi Mujer to escape. (Photo Jesús Rivas)

differences mean a substantial difference in the life of anacondas in the rest of its distribution? Green anacondas are mostly distributed in areas of ever-green forest, with little seasonality, deeper rivers of permanent water. Those are the traits that make studying anacondas in other places more difficult. How influential might they be in dictating a different ecology for the animals that live there?

As I have explained, anacondas that live in deeper rivers may face less predation, enjoy easier locomotion due to the lack of severe droughts, and encounter more food. All these are the traits that may limit growth for the an-imals (see Chapter 6). The strong seasonality is the reason it is easy (for me) to find females in the breeding season. Does it also make it easier for the males? In other words, in areas that lack seasonality, will many males find the same female at the same time? If not, there would not be breeding aggregations in the other parts of their distribution. They seem to make breeding balls in sea-sonal Cerrado (Juliana Terra, personal communication), but we do not know what the situation is in the less seasonal Amazons basin. There are no data or information reporting that this is so. Is this because they don't make breeding aggregations in flooded areas or is it because we haven't studied them there? Putting it all together, does what I have learned about anacondas in the past 25-plus years represent valid knowledge of the ecology of the snake, or does it represent just the particular conditions in the llanos?

I have been critical of North American herpetologists for their convenience-based approach to the study of mating in garter snakes. Garter snakes are probably the best-studied snake in the world as they are distrib-uted all over North America. Yet the only studies of mating in garter snakes come from hibernacula in the uppermost point of their distribution, where it is possible to find hundreds of snakes in a single day as they come out of hi-bernation. Thus, all the data about mating in garter snakes come from these locations, even though garter snakes are distributed all over North America, where they do not mate that way. So, the reported mating system of garter snakes may well be an exception to the rule and not representative of their true ecology. This is not all the different from what I have done. I have studied anacondas in the uppermost point of their distribution using a dry season that makes it easier for me to find them, but I don't know to what extent I can extrapolate this information to the rest of their distribution. Not unlike the North American herpetologists I have criticized!

This brought to my attention the issue of how it all happens. The llanos as we know it are only 10,000 to 15,000 years old. This is not enough time for evolution to do much. In other words, anacondas did not evolve in the current llanos. Does what I have learned about anaconda biology simply represent maladadaptions to the new habitat? Is this why they have such high mortality after birth? What is the habitat that anacondas are adapted to, if not the llanos? Is it the current Amazon the habitat I should be studying because that is where anacondas evolved? Where do anacondas come from?

When I was doing my Ph.D. at the University of Tennessee, a German colleague, Lutz Dirksen, was doing his dissertation with anacondas as well in Germany. His study was on systematizing and classification, so our work was very complementary and we corresponded regularly. Once Lutz asked me why there were not anacondas to the west of the Andes. This was a very interesting question because in western Colombia there is a huge swamp, Darien, that would be perfect for anacondas to thrive. It has caimans, capybaras, and turtles in abundance as well as other prey anacondas are known to consume. It is a huge swamp with plenty of good habitat—yet anacondas are not known to exist there. There is a lot less of South America to the west of the Andes than to the east, yet there is plenty of suitable habitat for them in many places, but they don't occur there. I didn't know the answer to Lutz's question, and I don't recall exactly what I told him. Since anacondas are strictly from lower elevations and the Andes is such a tall mountain range, it is easy to see how they wouldn't move there. But this can be said of other snakes, say boa constrictors, and we do find them on both sides of the Andes.

Wondering about Lutz's question and my other questions (Where do anacondas come from? Does what I know of anacondas in the llanos mean anything?) led me to wonder about the paleo-history of South America. Since ecology and evolution are the same discipline at different temporal scales, I figured I could not answer questions about anaconda ecology without tackling the evolutionary questions. Also, by doing so, I could know if my knowledge about anacondas in the llanos would apply to anacondas elsewhere.

The Birth of a River

A Rift on the Land

Going back in history 150 million years ago (mya), the current continents of South America and Africa were joined in a single mega-continent that also included current-day Australia and Antarctic. The northern part of this continent (modern-day South America and Africa) were drained to the west by a very large river that started roughly where the current Congo River starts and flowed out of what is currently western Ecuador. Approximately 110 mya South America separated from Africa and drifted west. The continent was drained by the paleo-Amazon, a river that was very much like the current Amazon but drained west. As South America drifted west, it collided with the Nazca plate in the Eastern Pacific. As both land masses moved against each other, the Nazca plate subsided under South America, pushing up the western border of the latter and creating the Andes, a 7,000-kilometer (4,350-mile) mountain range that stretches from northern Colombia to southern Chile and Argentina, the largest mountain range in the world.[124]

While there have been many studies on the geomorphology and geology of the area, most studies have ignored a rather crucial aspect: What happened to all the water of the mighty river? This is not a small omission. Anyone who has come face to face with the Amazon, or even the Orinoco, is humbled by the sheer immensity of it flow. Now, this paleo-Amazon was likely even larger than the current Amazon River. It drained all the landmass that is currently drained by the Amazon and Orinoco combined. These are the largest and the third largest rivers in the world in terms of volume. Furthermore, this was a time when the planet was a lot warmer. Tropical climates extended as far north as Montana. The ice caps were very small or nonexistent, and all that water was circulating in the planet, making it a much wetter planet than it is now. It follows that damming such a colossus of a river would have produced very important consequences in the hydrology of the continent, and likely the planet.

A Big Dam

The rise of the Andes would result in the eventual closing of the drainage of the paleo-Amazon into the Pacific Ocean. It is estimated that about 90 mya, the creation of the Andes started obstructing the flow of the paleo-Amazon into the Pacific.[124] It would have been a very slow process that at first did not have much impact. However, as the Andes rose, it would have resulted in a shallowing of the mouth of the paleo-Amazon, preventing it from draining at full volume. So, the first consequence of this process would have been that the river backed up at the mouth and overflowed into its natural floodplains. These were likely floodplains that flooded seasonally at first, but as the volume of the river increased, over time it would have started to flood the surrounding areas more permanently.

The mouth of the current Amazon is more than one hundred meters deep. If the paleo-Amazon had a comparable depth, given the slow rising of the Andes, the damming of the river would have been a very gradual process. Both of these, the slow pace of creation of the mountain range and the erosion that the running river would have caused, resulted in this being a very protracted process. Yet, as the mouth of the paleo-Amazon became shallower, there was more water to overflow the banks. Neighboring areas likely became permanently, or mostly, flooded, like the várzea forest and igapós we encounter today in several parts of South America. These are short canopy forests that spend most of the year flooded by a variable layer of water. Slowly but surely the flooding would convert the surrounding landscape into marshes, swamps, and other kinds of flooded habitat (Figure 9.1).

Sedimentological studies show that there was a great river that flowed over the Llanos/Magdalena basin and drained into the Caribbean from what is currently northern Venezuela and northeastern Colombia. It was very shallow and had a very broad riverbed.[125] Instead of a flowing shallow river, it was more like the overflowing of a continent that had lost most of its capacity to drain water into the Pacific Ocean and spilled over to the north into the current Caribbean Sea. This general flooding of the northern part of the continent was likely what led to the conditions for the evolution of the giant aquatic serpent found in eastern Colombia, *Titanoboa cerrejonensis*. The large extension of flooded tropical habitat had huge primary productivity, given all the sun, all the water, and nutrients from the rising Andes, and had an abundance of all sorts of vertebrates. This set the stage for some large boa to exploit the resources in the water and develop into the fabulous size it did.

The huge volume of water of the paleo-Amazon could not be expelled via the original mouth of the river as it became progressively shallower, and part of that water overflowed to the north. It is possible that some also flooded the lowlands between the southern Andes and the Brazilian shield via the Madre de Dios and current Pantanal.[126,127] So, all the lowlands of the continent would have become permanently, or seasonally, flooded, producing a landscape of marshes and swamps in most of the western Amazonia, probably extending as far east as Manaos. The main landmasses above the water would have been the incipient Andes ridge, the Guyana shield in southeastern Venezuela and northeastern Brazil, and the Brazilian shield in southeastern Brazil. Seasonality in the lowlands would have produced expansion and contraction of the water bodies, not unlike what we see currently in the llanos and Pantanal.[5,128]

A Mega-Wetland

Because geological movement is so slow, it is unclear when exactly the connection with the Pacific and the paleo-Amazon, the Ecuadorian portal, was severed. It is believed that about 33 mya there was no significant volume of water draining into the Pacific. At this point the continent was very moist because of all the water that could not be expelled into the Pacific. Yet, it would become a lot wetter soon. The rising of the Andes started trapping the moisture of the trade winds going into the Pacific. As the winds blew west, the Andes pushed them up. On rising, the winds would cool down and drop their moisture, producing substantial orogenic precipitation.[127,129] At this time South America continued receiving moisture through the Atlantic trade winds but the air masses that left the continent, after going over the Andes, were dry. This resulted in South America sequestering large amounts of global moisture.[130] As the Andes continue to rise, the continent would retain more and more moisture that was not draining anywhere. With the flat relief of the continent small changes in precipitation would have resulted in expansion and contraction of water bodies, producing vicariant events in terrestrial and aquatic ecosystems leading speciation pulses all over the continent. This probably explains South America's high diversity compared to other tropical landmasses. By 23 mya, a mega-wetland developed over most of western Amazonia, the Pebas system. Orogenic precipitation returning water back to the continent also washed substantial sediment from the Andes. Accumulation of this sediment over the years filled up the Pebas system. Approximately 7 mya, because of this sediment, the Amazon River (as we know it today) appeared, flowing in its current easterly direction (Figure 9.2).[131]

What does all this mean to the evolution of anacondas and to how relevant my data are for the big picture of anaconda ecology? It seems like the conditions in the paleohistory of the continent of constant flooding are not all that different from the conditions that anacondas encounter currently in the llanos. It also shows that there might have been areas of permanently deeper water and areas at the edges of these swamps where the bodies of water expanded and contracted seasonally. This is not that different from the current conditions in South America. Deeper oxbows and lagoons exist, often around the Amazon and deeper rivers. Also, at both the northern and southern borders of Amazonia there are large extensions of very seasonal habitats with strong wet and dry seasons (the llanos and Pantanal, respectively), not unlike those that bordered the original mega-wetlands in South America's paleohistory.

The Players

Given this scenario in ancestral South America, we can now try to figure out how it affected the evolution of its inhabitants. We have four extant genera of boas in South America.

Corallus is a group of strictly arboreal snakes with two species in South America: *C. enhydris* and *C. caninus*. They are both distributed through the lowlands of tropical South America. These snakes are limited to areas of continuum forest canopy as they are tree specialists. They have very slender bodies adapted for climbing, they have frontal-facing eyes, and they specialize in an avian diet (Figure 9.3). They are the only group of snakes that give eye-shine, as a consequence of their nocturnal habits.

There are two genera of generalist snakes, *Boa* and *Epicrates*, with a long habitat distribution. Both are widely distributed through the continent, with *Epicrates* being more specious. They seem equally at ease climbing trees, crawling, and swimming. They do not have obvious adaptations to any particular habitats.

There is one current genus of strictly aquatic snakes: *Eunectes*. These are the different species of anacondas we know. They have eyes and nostrils on top of the head, seemingly adapted to live in water and stalk prey on land. This genus contains the green anaconda (*E. murinus*), which is distributed all over South American in the Orinoco and Amazon basins, extending as far as southern Brazil and Paraguay. It can be found in rainforest, dry forest, the Cerrado, in the Southern Hemisphere and savanna habitats in the Northern Hemisphere (Venezuela and Colombia). This is the species that grows truly huge. There are three other species of smaller anacondas: (1) yellow anacondas (*E. notaeus*) (Figure 9.4), which live in Pantanal, Chaco, and neighboring areas in southern Brazil, Bolivia, Paraguay, northern Argentina, and Uruguay; (2) *E. beniensis*, which has been only recently described and is known to occur only in the Beni in Bolivia (Figure 9.5); and

(3) *E. deschauenseei*, which is supposed to live near the mouth of the Amazon in eastern Brazil and French Guyana.[132] However, this classification is based only on morphological data and there is no molecular study of the group, so it may contain cryptic species as well as duplicated ones. Although I am working in the area of molecular analysis of these groups to resolve some controversy, it seems clear that *E. notaeus* and *E. beniensis* are sister taxa. They are also probably highly related to *E. deschauenseei,* while *E. murinus* would be the outgroup within this clade. Green anacondas are then the larger ones, the ones we find in deeper bodies of water throughout the Amazon, while *E. notaeus* and *E. beniensis* are more of a savanna-dwelling anaconda, occupying more seasonal habitat. The big mystery is E. *deschauenseei*, which seems closely related to the *E. notaeus/beniensis* clade but is surrounded by *E. murinus* in its distribution and occupying habitat used by the larger clade.

The Tree

Looking at the relatedness of these groups, we find that *Boa* separated earlier from the rest of the clade some 58 to 70 mya. Between 43 and 55 mya *Corallus* split from the group. Then, approximately 29 to 33 mya, *Epicrates* separated from *Eunectes* (Figure 9.6).[137] What happened within *Eunectes*? When did the different clades separated? How and why? With our current knowledge and understanding of South America paleohistory, I am prepared to give an educated guess of what might have happened.

Because they are a strictly arboreal taxa, the appearance of *Corallus* must have been facilitated by the diversification of angiosperms that had appeared in the world relatively recently. *Corallus* needs continuous forest canopy, so their appearance required continuum forest and perhaps the presence of lianas. Likely large areas of marsh and swamps that disrupted the forest canopy would have been a barrier for them to cross. However, when the continent flooded toward the west, the northeastern Guyana shield and southwestern Brazilian shield would have maintained abundant forest for them to evolve.

On the other hand, the appearance of *Eunectes* was probably tied to the swamps created by the big dam. The earliest fossil of *Eunectes* known was from about 15 mya,[134] but we have already discussed how difficult it is for fossils to form, to begin with, and then how unlikely it is for them to be found. In flooded habitats bone would decompose quickly. Stagnant bodies

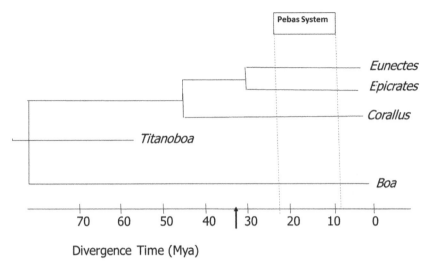

Figure 9.6 Phylogenetic tree of the boas in South America. Arrow indicates the time when the connection between the proto-Amazon and the Pacific Ocean was closed or was close enough that it led to substantial flooding of the continent.

of water (which anacondas like) often develop an acidic pH due to the leaching of tannins from decaying leaves. The calcium phosphate from the bone dissolves very quickly in an acidic pH, further reducing the chance that fossils will be formed. So, for this analysis I will ignore how recent the first *Eunectes* fossil is and instead rely on the molecular evidence. This line of evidence suggest that *Eunectes* lineage separated around 33 mya. This is exactly the time when the Ecuadorian portal closed and the flooding of the western Amazon became more dramatic. This was perhaps not a coincidence. The existence of large areas of flooded habitat, with great productivity, surrounding a few elevations that would have become islands, was probably a perfect scenario for some existing boas to adapt to aquatic living.

Now, the nature of this habitat was not like others we see today, at this scale. It was not a deep lagoon or a river. It was a huge area of relatively shallow habitat where the water was mostly stagnant. It represented water from the big river that flooded the northwestern part of the continent and very slowly spilled toward the current Caribbean. This swamp, with a high level of nutrients from the Andes sediment and high tropical productivity, would have easily been covered by macrophytes. Probably large areas of water hyacinth and other rooted aquatic plants grew, covering the water surface. Anacondas' coloration and morphology seem perfectly adapted to this habitat (Figure 9.7).

A tropical shallow body of water covered with vegetation would develop anoxic conditions very quickly. These waters were not hospitable for aquatic vertebrates that breathe under water (amphibians and fishes). Probably only very small fishes and amphibians could survive in these conditions. So, it is likely that anacondas evolved eating a diet of mostly terrestrial vertebrates and those aquatic ones that breathe air (such as turtles and caimans). This helps explain the troublesome lack of fish in their diet. Living in water and hunting terrestrial prey are what anacondas' superior eyes and nostrils suggest. This habitat was probably a scaled-up version of the same habitats we find currently in the floodable llanos during the wet season, although the seasonality was probably not as strong. Likely there was a seasonal variation of precipitation, but there would have always been depressions with enough water for snakes, so they didn't have to crawl across dry land to survive. This scenario would have led to the evolution of a very large aquatic snake like *E. murinus,* since the limitations that we have identified for large size were not there. Because there was a lot of water all the time, large snakes could hide their large bodies to stalk prey, they didn't need to crawl across dry land, locomotion carried little cost thanks to the water level, they could drag their prey down the water and avoid retaliation from prey relatives, abundant prey existed thanks to the high productivity, and there was little to no penalty for large size.

Toward the southern and eastern borders of the Pebas system there would have been flooded areas experiencing hyper-seasonal fluctuations of the water table, not unlike the current savannah of the llanos and Pantanal. With very strong seasonality, *Eunectes* faces the problems of having no or little water at some points in the year, so that would have resulted in a limit for large size in this margin of the Pebas system. I cannot identify, at this point, any zoogeographic barrier separating these lineages. A working hypothesis is that they would have been separated by their use of different habitats mostly, with some species specializing in deeper, more permanent water and others specializing in temporal flooding conditions. However, these conditions lasted long enough, and the area is large enough, that speciation would have occurred. When the Amazon River was born 7 mya, most of the South American basin became covered with continuum forest. The swamps and marshes of the Pebas system gave way to forest areas and flooded forest, displacing the savanna anaconda to the south where we currently find the

E. notaeus/beniensis clade. The swamp-dwelling anaconda (*E. murinus*) remained in the rivers and smaller marshlands in the Amazonia/Orinoco complex, colonizing the newly developed savanna toward the north end of the former Pebas system that eventually became the llanos.

It is uncertain how *E. deschauenseei* came about. A working hypothesis is that some of the smaller savanna anaconda were progressively displaced to the east (the higher elevation of the original continent) as the forest expanded from the west. The original savanna anaconda would have ended up engulfed by forest eventually and likely specializing in the shallower, savanna-like areas that still exist in eastern Amazonia. Because the information on *E. deschauenseei* is so limited, we don't know if it is able to interbreed with the surrounding *E. murinus*—in fact, we don't know much about their biology at all.

10

Conservation of Anacondas and Beyond

The Interface of Biology, Politics, and Economics

A time comes when silence is betrayal.

Dr. Martin Luther King, Jr.[135]

Solo le pido a Dios
Que el dolor no me sea indiferente
Que la reseca muerte no me encuentre
Vacío y solo sin haber hecho lo suficiente.

(I only ask from God,
that suffering may not be indifferent to me
That the dry out death does not find me
Empty and alone without having done enough.)

Lyrics of popular song "Solo le pido a Dios" by Leon Gieco
(Translated by JAR)

In March 2003, I took a break from pursuing the anaconda's life history. I had landed a wonderful opportunity to join two climber friends to go up the wall of Roraima's Prow, a *tepuy* (mesa) located between Venezuela, Guyana, and Brazil. The Prow is located in Guyana and had not been climbed success-fully to that date. It was an exciting adventure where I could dust off my old climbing skills, and it also gave me the opportunity to collect specimens of fauna and flora in the vertical part of the wall that, due to the difficult access, had not been sampled. It was the same kind of habitat where, as a sopho-more, I had the opportunity to collect specimens of toad (some species of *Oreophrynella* sp.) that had not yet been described. It all promised to be an exciting trip that involved trekking in the rainforest, deep contact with na-ture, and hanging out in a gorgeous patch of virgin forest. On arrival to the wall, two professional climbers did all the technical climbing work, leaving me to do the biological collection of plants and animals, mostly invertebrates.

A film crew from *National Geographic* and other colleagues came along to document the expedition. It was a most enjoyable time, hiking 10 hours a day, exploring the forest, sleeping in hammocks under a tarp that shielded us from the ever-present rain of the rainforest, and waking up to the calls of howler monkeys and forest birds. As we were working our way to the Prow, we occasionally had radio reception to keep us informed of what was happening in the world. One day we got the news. It was March 19, 2003. The United States had started another unnecessary war. While some people in the United States might have been fooled by the constant propaganda that the war was a national security need, I had a bit more of an international perspective and I could not buy it. It was clear to me, as it was clear to just about anybody outside the United States, that the only reason the United States was invading Iraq was to appropriate their oil.

Then the realization hit me. We were going in the noblest pursuit, seeking to explore virgin parts of the forest, hoping to describe new organisms to add to our knowledge of the natural world. We were making all efforts to do the exploration, and we also brought a TV crew along so everybody could be part of it. We were doing good science, exploring new areas, documenting the natural world, and bridging the gap between the scientific world and the lay audience. We were covering all the bases—and all for what? All over the world there are thousands of scientists working hard to make this planet a better place. There are millions of activists doing their best to address the environmental crisis. What good is all that if the few people who hold all the power continue to drive the planet into the dumpster, led by their ambition and greed?

It was nothing but a looting undertaking to put Iraqi oil fields at the service of transnational corporations at the cost of countless innocent lives, not to mention the environmental degradation associated with it. Was it any different than barbarian hordes of the old days descending on unsuspecting communities, burning and looting all they found in their path? Where was the international community? Where was the minimum respect for laws and basic decorum? Where were the pretenses of having a civilization? What was the point of preserving the natural world if the most powerful nation in the world could give in to the greed of the corporations that rule it and engage in such a blatant atrocity? What is the point of working in environmental conservation? What good would it be for us to document more of the South

American biodiversity? What good can it be to find, say, a new species of moth, or new variety of beetle or frog? Can people watch a beautiful nature documentary on TV and switch the channel to see bombs killing innocent children and not see the contradiction? How many endangered species were going to be killed by cluster bombs in the countryside of Iraq for the sole purpose of satisfying the unquenchable greed of transnational corporations? Why bother doing research and learning about animals and wildlife? What does it matter to develop conservation plans that will protect biodiversity if we are all hostage to the greed of a few people who do what they want because they have the resources, political clout, and media complicity? How can we bury our head in the sand and ignore the fact that our research will not go anywhere so long as the fundamental workings of society are all going in the wrong direction?

What will be next? Will no one stand in their way? Do these people know no boundaries? Do they know no shame? With 55% of the oil reserves in the word, Venezuela has not only the largest oil deposit in the world but also more oil than the rest of the world combined. What is standing in the way of Venezuela, a mega-diverse country, from becoming the next target for an oil war? Maybe they will spare Venezuela because of all its mega-diversity and conservation value. Like hell they will! The same people who did not blink before ordering the slaughter of countless Iraqis would not really care much about the environmental damage that invading Venezuela may produce. Neither will they care about Venezuelan lives. And 17 years after the Iraq invasion the military–industrial complex is getting ready to repeat the same number in Venezuela, barely even trying to disguise their actions under the ironic veneer of "protecting human rights."

Conservation is a relatively new branch of biology that has one foot firmly set in science and the other firmly set in economics and politics. However, most of the conservation work we do is strongly biased toward the biological side. This is understandable because when I say "we" I mean the scientific community, and doing biology is what we know. But in the next section I will take a trip outside of biology to put conservation into context.

Latin America: An Economic Perspective

Those who have been out in the rural areas of Latin America have had the opportunity to enjoy great and beautiful landscapes and pristine natural ecosystems, but they have also had the less pleasurable opportunity of seeing how local people live, their economic situation, and their limitations and struggles. It becomes immediately obvious that there is no amount of education, policing, or enforcement that can really prevent them from using the natural resources around them to survive.[136] In fact, it would not even be humane to expect them to do so. It is also evident that these people do not have many ways of obtaining money but are still subject to living in a money-driven system and are very easily persuaded with little economic incentive to use nature in an unsustainable manner. The sale of wildlife as pets or for their parts is a common example of the latter.[137] Whether they use nature unsustainably of their own accord, out of lack of education and environmental awareness (i.e., subsistence overhunting), or whether they are encouraged to do so by external pressures (i.e., encouraged by an external/international market that produces the economic pressure[138]), it is clear that the abject poverty in the rural areas is the main conservation problem, and no conservation program can succeed if it does not address this in a direct and bold manner.

Not Just Pure Science

A lot of people are surprised to see how politically involved I am. Somehow they expect that I would be so focused on anacondas that there would be nothing else on my mind. In a way, this would be a contradiction of some sort. I cannot conceive how a person can be passionate about something and completely apathetic and dispassionate about other things that are so important. It would require a great level of mental dissociation. How a person can be so passionate about, say, ethical treatment of animals, and completely ignore other ethical aspects involving wholesale destruction of nature, or horrible human rights violations? I guess it is because I am Venezuelan. And I don't mean only that I am a hot-blooded Latino, which I am, but also living in Venezuela granted me opportunities, for the lack of a better word, I would not have had otherwise. Part of the reason for my political awakening was the fact that I have had the opportunity, for the lack of a better word, of witnessing five *coups d'état* and a social explosion that occurred in a country when I happened to be living there. I was in Ecuador teaching a course on tropical ecology during the coup that overthrew Jamil Mahuad in 2000, and

I also happened to be there in 2005, teaching a similar course, when Lucio Gutierrez was ousted. I was in Venezuela during the coup against President Hugo Chávez on April 11, 2002, and also was lucky enough to have been part of the popular uprising that took the country back to restore democracy on the 13th. That's four. The other *coup d'état* that I witnessed was the 2000 election in the United States that, I know, time will eventually regard as a rupture of the constitutional line in the United States. Finally, I was in Caracas on February 27, 1989, when the country was shaken by an all-out social explosion known as *El Caracazo*. If I could choose, I would rather experience five more *coups d'état* than witness another social explosion!

In the next section, first I discuss the technical aspects about anaconda management that can be used for conservation of anacondas and conservation of biodiversity in general. But I cannot in good conscience regard these as conservation activities. These are issues about management, not conservation. A lot of people equate management with conservation. As I have argued in the past, and will argue some more here, they are certainly not the same.[139] Then I address holistically the problem of conservation in Latin America and beyond.

Wildlife Management and Conservation in Venezuela

Venezuela withstood the early onset of the economic crisis better than other Latin American countries due to the fact that all the country's oil reserves belong to the nation, not private companies. So, a lot of the oil revenue was used to subsidize the needs of the people. However, from 1982 to 1998 the country transitioned toward a neoliberal approach (see later in the chapter for more details), and there was a slow but consistent decline in the economy that affected the lifestyle of the people and, ultimately, the environment and wildlife. As the economy of the country worsened and the wages of the local people fell well below the minimum necessary to survive, people started using resources they would have disregarded otherwise. For instance, in the past capybara meat was eaten only in the week before Easter; it was a tradition in some regions of the country. Now, illegal hunting of capybaras occurs throughout the year, as people have resorted to eating capybara as a staple of their diet. Traditionally, the cattle ranch that produced most of the country's capybara meat was El Frío. For more than 30 years, El Frío sustained an estimated population of roughly 30,000 capybaras, of which 10,000 were harvested every year.[140] However, in 1986 I participated in a survey of El Frío capybaras and the population was only slightly above 4,000! Later surveys of the area indicate a further decrease in the population, and poaching has been acknowledged as the leading cause of the population crash. Similar cases of significant poaching have occurred with other species, including white-tailed deer, caiman, iguanas, side-necked turtles, and peccaries. This trend is, not surprisingly, expected to continue and extend to other species as the poverty of the country worsens.

Pressures on Anaconda Populations

Although use of the green anaconda is forbidden by the government, anacondas have been harvested illegally. When I started studying anacondas in 1992, I learned from some local people that the tanneries were paying Bs.1,000 ($16.67) per meter of skin. This is a significant amount for a worker who made approximately $3.50 a day. The skin of an average anaconda would provide more income than a whole week of work! In other countries of South America anacondas have been harvested legally in low-profile programs. In Guyana anacondas are harvested opportunistically by fishermen and sold to local tanners.[2] In Argentina *Eunectes notaeus* has been harvested legally[141] and in Paraguay illegally (Patricio Micucci, personal communication). For a few years Bolivia has been considering a program for sustainable use of *Eunectes murinus* but has not done so yet (James Aparicio and Luis Pacheco, personal communication). It is well known that artisanal use of anaconda skin and byproducts is practiced around its area of distribution, but its impact is likely small since these are not high-impact operations (Figure 10.1).

The rational use of wildlife has been used as an alternative to its destruction. For example, several populations of crocodilians that have been seriously threatened are now recovering due to effective harvesting practices (see reference 146 for a review). However, there is a thin line between management that is used as a conservation tool and management that represents just another way to use nature to make money with little or no help for conservation. Is the goal of the management to conserve nature, or is it business as usual? And if this is just another business, is it at least a sustainable one?[139] In the next section I review management methods and how they can be applied for the genus *Eunectes* as well as the socioeconomic implications of management.

Wildlife Management as a Tool for Conservation: Harvesting versus Farming

Farming

The most common methods of extractive wildlife management are farming, harvesting, or a combination of both. In a farming model, animals are kept in captivity, and all their needs are provided for by the keepers. This is a relatively expensive activity and is preferable for animals that have high growth rates and low maintenance expenses and can be housed in high densities. Farming anacondas in a closed system is unlikely to be successful. The cost of facilities and maintenance would probably be prohibitively high. It is unlikely to be cost effective to maintain a species that eats meat and that takes several years to reach adulthood, and where females will not breed every year but every other year at best.[15]

However, the possibility of an open farm system exists. Large pregnant females can be found along the riverbanks,[15,4] caught, kept in captivity, and released after parturition. Due to their high fertility, a large number of individuals can be produced in short-term farming or for the pet trade. Neonates have a high natural mortality in the field,[4,14,15,63] and protecting them in captivity and releasing some later would result in no net loss for the population. Neonates can have a relatively fast growth rate,[4,143] and, after a short time of farming, can provide excellent, scar-free, small-scaled skins that would have a high market value. In addition, young individuals have a sharper pattern and more attractive skin.

Anacondas do not make good pets. They quickly outgrow their cages and become a risk to other pets and even people. They have a feisty temperament and seldom become an easy (or safe) animal to handle. Although it is possible to tame them, it requires a long time and lots of handling, and they may revert to their untamed ways if the handling stops for a relatively short time. They also release an aversive, very fetid musk when disturbed. However, due to the charisma of the animal, anacondas are popular in the pet trade (approximately $250/neonate, retail). There are some anaconda lovers who would violently disagree with the last few statements. But even they will agree that anacondas are not a pet for just anyone to have. Devoted snake-keepers manage to tame them, resulting in very large docile pets, but it's just not a pet for the novice or the impulse pet buyer. The illegal import of live reptiles for the pet trade is a growing market in the United States.[144] Because most reptiles can survive for many hours without water or food, the animals can be smuggled into the country in many ways. This market is very hard to control and the number of animals being extracted is difficult to quantify.[144] Thus, a legal source of neonates that come from a sustainable system could be a way to promote protection of the wild population. There are breeders in the United States and Europe, but their production capacity is relatively limited compared with the larger supply that an open farm can produce. In other words, an open farm could compete favorably with these operations and bring much needed revenue to developing countries.

However, a supply of legal anacondas from the field could increase the demand for anacondas, and this demand would create pressures for the unsustainable use of natural populations. So, this practice is not without risk. Furthermore, importing live animals leads to challenging ethical issues regarding the welfare of the animals, as they might end up in the hands of novice pet owners who will not keep them in optimal conditions. For larger reptiles, there is always the question of what to do when the animal outgrows its facility. Many adult snakes exceed legal size limits dictated by urban areas. What do you do with illegal pets? Frequently the animal is turned loose in an exotic environment where it will, at best, die in a short time from exposure or starvation—although sometimes it survives and reproduces, causing further problems as an exotic invader in a foreign ecosystem (Figure 10.2).[145]

Anacondas have been banned as pets in some of the southern US states due to the concern that they can establish feral populations. However, among the snakes that can produce feral populations in the United States, green anacondas are about the least likely, in my opinion, because they prefer warmer temperatures. In all the telemetry studies I have done, I have never found an anaconda in water any colder than 22°C (71.6°F). For instance, I believe that yellow anacondas have a stronger chance of becoming established in the Everglades because it is similar to their native climate. Since the area where they live does experience brief periods of freezing, like the Everglades, and since they are live bearers, which allows them to endure cold snaps better than egg layers, they may even be a strong competition for the Indian pythons that have been living there for more than a decade.

Closed farming does not represent a threat to the wild population since only a few animals are originally collected from the wild, and if the project fails, only the animals that were in the farm are in jeopardy. Also, due to the localized nature of the activity, it is potentially easy to monitor and enforce the existing regulations. Closed farming, however, is an activity that benefits the few people working on the farm and does not require pristine habitat.

Therefore, farming does not put any pressure on communities to protect nature, nor does it produce many jobs for local people. Consequently, farming has a rather modest impact on the economy.[146] Thus, although sustainable, farming would not be a constructive conservation method but, rather, would represent business as usual. This is simply one that uses a species of wildlife. A system of open farming, on the other hand, has more potential of being used as a conservation tool as it requires a natural environment where the animals live[151] and provides benefits for more workers, but as we will see later, it is not conservation by itself.

Harvesting Anacondas: Pros and Cons

On the other end of the spectrum is harvesting or cropping. In a cropping system, animals are harvested from the wild; thus, a direct link exists between the economic activity and the conservation of the species and its habitats. The economic incentives the locals receive are directly linked to the habitat, producing clear reasons for them to protect and not overexploit natural areas that provide the resource under exploitation. Thus, cropping has real potential to be used as a conservation tool, but like open farming, it is not conservation by itself (see later for more details). This activity is better for animals that occur in high densities and are easy to find and catch. It requires a much lower overhead than farming since the only investment involves finding and catching the animals that are going to be harvested. However, due to the more extensive nature of the harvest, it has a much greater potential to have a detrimental effect on the natural population if it is overdone. Monitoring and controlling the harvesting activities are a great priority, but it can be very expensive and technically, as well as practically, challenging.

Population Estimates

A good management plan for wildlife includes a good understanding of basic life history. Even modest success at wildlife management requires some knowledge of the population parameters, demography, and the maximum sustainable yield a population can support. Experimental programs based on indexes of abundance are also possible, but they need to be assessed very carefully to prevent overexploitation. The main population parameters are abundance, rate of increase, fecundity, mortality, recruitment, and dispersal. Ideally, population size followed by the intrinsic rate of increase of the population should be determined. These statistics will enable us to calculate the maximum sustainable yield (MSY), which is the maximum number of individuals that can be removed from the population while keeping the population essentially constant.[147]

The first problem encountered when attempting to harvest anacondas is their secretive nature. To harvest a population rationally, we must be able to count how many animals there are in order to propose a sustainable harvest rate. If we don't know the total number of animals available, the alternative is to have some estimate of the population size in the form of an index of relative abundance (e.g., number of snakes seen per kilometer of road). This way we can make an educated guess about the MSY and refine it by monitoring its impact on the population by changes in the index of abundance. In this way we can detect any problem and fix it in a timely fashion.[147] For instance, if we note a substantial decline in a population that is being harvested, we can lower the harvest rate until the decline stops.

To date, we do not have any of these surveying tools with respect to anacondas. To estimate the abundance of the population necessitates long-term mark-and-recapture studies that are too time-consuming to apply to the large-scale management of the species. We do not have any good index of relative abundance either. Due to their secretive nature, none of the traditional methods of counting by transects can be applied in a simple manner for anacondas. A possible method for developing an index of relative abundance for anacondas may be by using the sighting of pregnant females at the riverbanks or edges of roads. Because pregnant females bask frequently along riverbanks and near the roads, it might be possible to develop an index of relative abundance using the frequency of sightings related to distance and duration of surveying to develop an index of relative abundance. However, since this method is based on the productive sector of the population (reproductive females), by the time it changes (in the case of overexploitation) serious damage to the population may have already occurred. Since we have no field-based method to monitor the impact of the program, harvesting of anacondas would be very difficult to monitor and there will always be the risk of overharvesting. On the other hand, anacondas have high fecundity, so we expect them to be fairly resilient if the harvesting is stopped in timely fashion.

Cropping the Animals

Capturing the animals for harvest offers another challenge, which is the number of hours needed to find only a few animals. Paying a crew to look for anacondas might not be cost-effective considering the low frequency of capture that I encountered (see Chapter 4).[4] One alternative strategy to overcome the low encounter rate with anacondas is to put together a crew that harvests other species as well, such as caimans (*Caiman crocodilus*), turtles (*Podocnemis* spp.), iguanas (*Iguana iguana*), and tegus (*Tupinambis teguixin*).[148] All of these reptiles occur in relatively high density and are potentially manageable. However, to implement sustainable management, there is much that has to be learned about these species, as well as improving the organizational skills of governmental agencies to manage all of these species correctly.

Other problems possibly encountered with anaconda harvesting are related to sexual size dimorphism and the enforcement of the harvest. Hunters involved in wildlife harvest typically tend to target the largest individuals first, which are usually males in many game species, because they provide more skin or meat. In polygynous species this is potentially sustainable since most of the matings are performed by a few males, and there is a theoretical surplus of males that are not breeding at a given time. In anacondas, however, since they are polyandrous,[4,15,19] harvesting the largest individuals is potentially devastating, as harvesting larger animals will involve harvesting females due to the female-biased sexual size dimorphism.[18,19] Also, larger females make the largest contribution to the population. Females larger than 340 centimeters (11.2 feet) are responsible for 59.5% of the new offspring every year, and females larger than 300 centimeters (9.8 feet) contribute to 74.8% of the total number of newborns in every generation.[15] In other words, any harvesting of large females would dramatically affect the population numbers, making cropping extremely risky to implement.

It could be argued that harvesting males is a more feasible alternative as they are more abundant, they are easier to find, and they can be gathered in greater numbers in the breeding aggregations.[4,15,19] Having smaller size and feeding on less dangerous prey, males tend to have better skins with fewer scars, thus increasing the quality of the product. If the program is created in a manner to encourage the collection of smaller animals, the odds of success are better, since they are more likely to be males and thus will have skins with fewer wounds and smaller scales.[2,4,15] However, even this alternative might be unfeasible given the practical problems mentioned earlier. Furthermore, since females that are courted by several males have higher reproductive success,[15] the quota of males for the harvest would have to be assessed very carefully.

Commercial use of large snakes is practiced in Sumatra, where reticulated pythons (*Python reticulatus*), blood pythons (*Python brongersmai*), and short-tailed pythons (*Python curtus*) are harvested serendipitously near plantations and villages. The snakes are kept alive in bags and taken to slaughterhouses, where the animals are processed.[149] This method targets mostly males due to their higher mobility and produces a variable rate of harvest that changes with snake abundance. Given the nature of this kind of harvesting, in which the hunters are not going out just to catch snakes, this method of hunting has the potential to be self-regulating. A drop in the population will produce a lower encounter rate that will result in a lower harvest. Given the cryptic nature of these species, it is unlikely that they can be hunted out or driven to extinction by harvesting. In the cases of *P. curtus* and *P. brongersmai*, the animals feed heavily on rats in the plantations and are thus also perceived as performing a pest-control role, which helps the survival of local populations.

A similar method is used in Guyana with green anacondas. Fishermen gather snakes opportunistically and keep them in bags to take to the tanners, where the snakes are killed for their skins. If the tanner considers an individual snake to be inappropriate for the market (too small, too many scars, too large), the animal may be turned loose. Although this has the potential to disrupt local genetic structures, this risk might not be very high since the tanneries are generally near the places where the animals are caught. Similar to the python harvest, this method seems to be sustainable since this low rate of cropping is not expected to threaten the population and the fact that people do not go out purposefully to look for the animals. However, any harvest based on encounter rate with people must still be regulated by a quota. With increases in human density or increases in the prices of the skin, the harvest rate could dramatically increase and eventually reach a level that might not be sustainable. Alternatively, if the country suffers a great downturn, the local people might feel compelled to hunt the animals beyond what is expected in the regular scenario.

Management of Anacondas

Anacondas and other boids are listed in Appendix II of the Convention for the International Trade of Endangered Species of Wild Fauna and Flora (CITES). This means that they cannot be the subject of commercial trade unless permits from the local government are obtained. In Venezuela, anacondas are still relatively abundant due to the large expanses of wetland habitat that lack human development and are relatively undisturbed.[5] There is no legal commercial trade of anacondas in the country; however, there is an illegal market for the skins. It is possible to find someone who will sell you an anaconda belt in this town or another. Due to the low impact of this activity, though, the pressure on the population is not too high and, at the moment, does not constitute a threat to the population.

The flesh of the anaconda, although edible, is not preferred by the local, or indigenous, people, and anacondas are not killed for it. Other than the skin, the only product of the anaconda that people seek (and more so than the skin) is the fat. Anaconda fat, melted under the sun in a closed container or on a fire, is considered a remedy for throat problems, asthma, and other respiratory problems. It also has been suggested that other anaconda products are used in homeopathic medicine to heal asthma and respiratory afflictions. As I was told by a practitioner of homeopathic medicine, the anaconda has the essence of asphyxia in it due to the way it kills its prey; according to homeopathic principles, giving something of the anaconda to someone with respiratory problems triggers the body to respond against asphyxia, thus curing asthma and other respiratory problems. Be that as it may, at present the demand for these medicines is not very high and does not produce economic pressure for its management.

In Venezuela selling anaconda skins is illegal and troublesome for the local people, so most people who do not live close to illegal tanneries do not engage in this activity. The main reason that local people kill anacondas is because they fear and dislike them so much that they will kill them on sight. Arguments that anacondas eat poultry, livestock, pets, or even people are often used to justify killing the snake (Figure 10.3). The truth is that people traditionally dislike and kill snakes even when they are nowhere near any of their livestock or houses. I observed straight, long scars or wounds on some live animals that I studied that could only have been made by a machete. This was especially true in the ranches that offer less protection to wildlife.

Habitat degradation in the llanos has not yet been a serious problem, since much of the land management for the cattle involves increasing the amount of land that contains water for a longer time. There is little farming, so agrochemicals are not a problem in general.[5] Much of the production is based on extensive cattle ranching, and the impact of this extensive cattle ranching on wildlife is much lower than the impact found in other countries where cattle are kept in higher densities. However, old-fashioned ranching practices involve cutting the gallery forests to ease the handling of the cows (which often hide in the forest and become feral) and to allow easy access for the cattle to water in dry season. This also removes habitat for the large cats that may prey on the cattle. Federal laws prohibit gallery forests from being cut up to 50 meters (164 feet) from the river, on both sides, but this regulation is seldom enforced. Deforestation in the llanos was not an important trend in the past, but it increased dramatically in the late 1990s, with an unsettling leniency from government authorities. The riverbanks often develop caves that are supported by the roots of the trees in the forest; frequently these caves are used by anacondas to hide and spend the dry season. In the treeless savanna, anacondas have fewer places to hide and protect themselves from extreme drought. This might be very significant in atypical years where the anacondas may be exposed to extreme heat or droughts.[4,15] The caves found in the segments of the rivers without forest are considerably less abundant and smaller than the caves found in other areas because without the roots, the river erodes and destroys the caves. Cutting of the gallery forest represents a direct threat to the anacondas' welfare. Of course, this threat to anacondas is in addition to the obvious effects that deforestation has on the populations of prey species and other components of the ecosystems, including all the forest-dwelling species.

Combining Science and Economy in Local Management

The use of management as a method to incorporate anacondas into economic development is not easy, and much more research is needed. Harvesting males and farming of neonates are possible alternatives that can be explored. However, both of these possibilities involve many practical problems as well as ethical issues that cannot be ignored. Killing animals for human comfort and leisure is a theme of heated debate on several levels between those concerned with animal welfare and those who manage wildlife for profit.[150] Changes in fashion or supply of other skins around the world can dramatically affect the demand, which affects prices paid for the animal products, along with the faith in conservation measures based on it.[146] I saw many ranchers in Venezuela make large investments in infrastructure to raise spectacled caimans for the skin market, but when the Florida alligator farms came online, producing lots of skins of higher quality, all the caiman farmers went out of business. I pity the biologist who tries to talk these ranchers into investing in wildlife management after they lost so much money in something that was supposed to be a safe investment!

New regulations adopted by the international community regarding the import of exotic wildlife, either in the name of conservation or in the name of animal welfare, can further limit the market and jeopardize all the investment made by the producers. Fluctuations in European fashion can make a product highly prized or worthless. Basing long-term conservation policies on the flaky fashion industry seems foolhardy at best.

Many countries may try to resort to their wildlife to solve economic crises. In Venezuela my research and recommendations managed to stop plans for harvesting anacondas since the oil wealth of the country relieved part of the pressure on the economy. In general, anacondas are at little risk of harvesting or poaching due to the problems mentioned above and the scarce overlap of humans and anacondas. However, generalized poverty in the areas, increasing human encroachment, and political and economic turmoil threaten the species, and these threats are expected to increase (Figure 10.4).

In my opinion, the clearest and least controversial benefit that local communities can gain from anacondas is the lure of anacondas, as "charismatic mega-fauna," present for ecotourism. The llanos have a tremendous and unrealized potential for ecotourism due to the large abundance and diversity of wildlife, comparable to the diversity of the rainforest.[151] Unlike the rainforest, in the vast savannas of the llanos the animals can be readily spotted and appreciated due to the lack of trees and the forest's naturally patchy distribution. Recycling of the profit produced by ecotourism into the local community in terms of jobs, education, and welfare would be vital for ecotourism to succeed as a conservation tool. The ranch where I studied anacondas provides a good example. When I arrived there in 1992, most people despised anacondas and saw no benefit to having them around. However, over the years, all the local people have seen how anacondas bring large numbers of tourists who come to the ranch just to see them. Filmmakers have brought substantial attention to the anacondas, which has produced an important economic incentive for the local economy (Figure 10.5). These tourists give generous tips to tour guides and generate jobs for cooks and the whole tourist industry. Other neighboring ranches have jumped in and started their own smaller operations with ecotourism-oriented activities. This halted the killing of anacondas on sight since now the people know that having an anaconda on their property can potentially represent money from the tourist industry. So, anacondas are providing resources for the people, which is a reason for them not to kill anacondas and to maintain their habitats (Figure 10.6).

Becoming a Radical: Avoiding Tylenol Conservation

As I write a chapter about conservation of anacondas, I need to make sure I do not mislead the reader into a false idea of conservation based on pure biology because that has been, in my opinion, one reason for the failure of the discipline of conservation biology. There are more and more conservation biologists every day. The discipline of conservation biology has grown by leaps and bounds, but the goals of conservation biology are every day farther and farther away. In my mind the jury is in: The discipline has failed.

Imagine that you hired a plumber to fix a faucet that leaks. When the plumber is done, the faucet is streaming a torrent of water that can't be stopped, all other faucets leak, all the drains of the house are clogged up, and the toilets back up when you flush them. Wouldn't you think this plumber is a failure? Yet, this is not a bad analogy to the state of diversity conservation when the discipline of conservation biology started and where it is now. After a few decades and great "progress" in the discipline, we are farther from success than we were decades ago. This is a fact that most conservation biologists seem to be oblivious to. And I don't mean that conservation biologists are responsible for the problems, but clearly the path that we have been taking does not get us to where we want to be.

I can confidently say that I am a radical in these matters, and not the superficial definition of radical but the true meaning. In today's media speech, the word "radical" is used for someone who is violent or who pursues his or her goals using violent means. That is not the true meaning of radical. Radical comes from the Latin root *radix*, which mean "roots." When I say I am a radical, it is because I truly believe that the solutions need to address the roots of the problems. Nothing will be accomplished by addressing the consequences and leaving the cause of the problems intact.

This is important when we talk about conservation. Conservation problems are many and diverse, but they all have one thing in common: They always have human origins. Nature, plants, animals, and ecosystems in general were doing great without us. The bottom line is that whatever the conservation problem, nature is not the problem; people are. However, most conservation research is oriented to manage nature. Wildlife management, forest conservation management, management of fisheries, and so on always deal primarily with managing the natural part of the problem, the part that did not have a problem to begin with!

When one has a problem, one can go to the root of it and eliminate the cause, or one can offer halfway solutions and ameliorate the problem without ever solving it. Imagine somebody who has a toothache. There is probably a very good reason for it, likely a cavity. One can try to seek the solution of the problem or one can take a painkiller. This painkiller is cheaper, easy to apply, and less painful than a dentist visit, but it will not solve the problem; it will just let us get by. With time, the painkiller will not be enough to mitigate the problem, and we will have to take higher doses of painkillers or use a more powerful drug that can numb the pain. In the meantime, the cavity keeps getting worse. Treating it early could be done with a dental appointment costing a minor amount of money and pain. Using the painkiller can pretty much guarantee that the problem will get worse, it will cost more money to fix, and it will produce far more pain.

The reader can see now the problems of our Tylenol approach to conservation. While we have spent lots of effort and resources on sophisticated research related to management of habitats and nature, we have given very little attention to the cause of the problem: managing the people. No matter how much progress we make in managing the consequences, the problem only gets worse because we have not addressed the causes.

I am not objecting to the idea that we should learn about nature, genetic flow, and habitat as well as resource management. What I am trying to say is that, just like taking a painkiller for a toothache, we should not expect to solve the problem by managing nature if we do not also make a substantial effort to manage people. Unfortunately, managing people is not what we were trained to do, and most scientists are reluctant to get involved in this kind of activities and often choose to stop short of doing what needs to be done.

Now, what does managing people involve? There are three basic ways to manage people that have been proven very efficient over the years: religion, education, and politics. Religion is well known to be a powerful tool to manage people and modify the behavior of the masses. However, most of us are still reluctant to recommend that religion be used as a conservation tool. I feel that it is similar to the ring of great power in *The Lord of the Rings*. In Tolkien's story the ring had so much power that nobody dared to wield it. Even the best-intentioned and best-prepared character, Gandalf, did not dare to wield the power of the ring because he knew he could be lured by its power away from his original goals. This is the way I feel about including religion as a tool for conservation. It has way too much power and I fear what it might lead to. However, liberation theology is an important movement throughout Latin America. It directly advocates Christianity to help the poor and the working class. Looking at the worldwide political landscape, Pope Francis has become the strongest and most unambiguous conservation advocate in the world!

Another tool for managing people, and one that most scientists feel comfortable with, is education. Most scientists, either by choice or obligation, are involved in teaching in academic institutions; however, that is not really producing the result it should. At the university level academicians might teach graduate students or undergraduate students. At the graduate level, I think that the universities, even conservation programs, select students based not on their potential for effective conservation but rather on their potential to do basic research, just like the people who do the selection, without really looking at the real needs for conservation.[152] Furthermore, at the undergraduate level, I also feel that academicians are missing the main target for education. College students are pretty much people who already believe in the conservation cause, a bit like preaching to the converted. The problem with conservation is not lack of knowledge but lack of caring. To illustrate this point I cite two examples. The first relates an event that took place in Panama, where a fellow ecologist was conducting some field research at the Smithsonian Tropical Research Institute. She and other researchers stopped at a restaurant by the coast where the menu included, among other things, turtle soup. Most of the researchers ordered the soup, but my friend (a native Spanish speaker) asked the waiter what kind of turtle it was. He responded matter-of-factly, "Green sea turtle." My friend translated this to her peers, but to her surprise they showed little reaction. The most concerned stated that as long as it was not a female, it was OK to eat. Of course, they didn't know if the turtle was a male or a female, but anyone with a minimum knowledge of turtle biology would know that because males are difficult to find, the animals most likely to be caught are females when they come to breed on the beaches. So, the turtle in the soup was more than likely a female, and a breeding one. Yet, the researchers then proceeded to enjoy their meal.

In contrast to this is an event I observed while working in an aquaculture farm in the foothills of central Venezuela. A medium-sized caiman had invaded one of the ponds and consumed a sizable amount of the fish being farmed. When the intruder was finally caught, I watched with surprise as one of the workers begged me not to hurt the animal, encouraging me to relocate it in a deep river nearby from which the caiman would most likely not return.

The researchers in Panama were postdoctoral and Ph.D. students in tropical biology and behavioral ecology. The worker at the Venezuelan farm had at most a first- or second-grade education and was raised in an area with little conservation awareness.

Perhaps these are not examples of any particular trend, but they are two extreme points that illustrate a more transcendental issue. Lack of knowledge is not the problem. It is certainly not the problem when lumber companies decide to clear-cut a large area of land in the old growth forests of western Canada or the Brazilian rainforest. Insufficient education in conservation issues is not the missing ingredient when large industrial companies decide to build factories in developing countries, where the local environmental regulations are not well enforced, if they exist at all. Lack of knowledge was not the problem when British Petroleum and its subsidiaries decided to save money on the seal to prevent fluids from leaking that resulted in the explosion of the Deep Horizon well, the largest oil spill in the world. It is clear that for the most part the real cause of environmental irresponsibility is not a lack of knowledge but a lack of caring. It's not information that will make a real difference in our environmental attitudes and behavior but, instead, feelings, since love cannot be bought, corrupted, or intentionally ignored the way knowledge can. We will not protect what we do not love.[153] Only those people who feel strongly about the environment and have established some bonds of respect and awe for nature are going to fight to protect it.[153]

Now, how can we teach the people to care? We know that teaching adults is good, but it can only go so far. Clearly teaching children is a more productive activity in the greater goal of creating conscious citizens, but who will do it? Unfortunately, although academics may agree with me, teaching children does not give tenure to university professors, and they do not see themselves bound to do this activity. They believe that this is the job of primary school teachers—poorly prepared, worse paid, and overworked primary school teachers, that is. To add insult to injury, any attempts to encourage schoolteachers to do this meets with a lot of resistance since they are so worried about standardized testing that they do not want to devote any time to other activities,

The last tool to manage people is politics. Politics and politicians can go a long way in changing people's behavior. Most academicians are willing to get involved in conservation *policies* but will run for the hills if you mention the word *politics*. In many places we have done superb conservation research and recommended great policies for conservation—but they were never implemented. In many case, it is, like Whitten and Mackinnon[160] pointed out, a bit of displacement behavior: We feel we are doing something good, but it is not really producing any results. It is not unlike a person who is frustrated at work because the boss mistreats him and comes home and goes for a run, or a dog that is intimidated by another one and engages in compulsive scratching. It does not help the problem at hand but distracts us from the uncomfortable situation. We feel a lot better talking about policies that we can back up with data and shy away from politics, which is the one thing that will actually make any difference. I really believe that we need to go the extra mile if we want conservation to be effective. We cannot hope that somebody will pick our policies and technical suggestions from complicated research papers and put them into action. Now and then, when I discuss the thesis project with a prospective graduate student, the student sometimes says: "I really want to do something good with my research. I want to make a difference for conservation." In these cases, I bite my tongue and don't say: "Run for office!" The people who have the knowledge on conservation do not have the interest in politics, and the people who are politically savvy do not know or care about conservation. Who is going to do it?

I know what you are thinking: "Politicians only care about votes." I won't argue with you, but this is good news: Politicians do care about votes! If voters are asking them to support "green" policies and implement environmentally sound policies, they will do it. We will not get conservation policies in place until we get the politics to turn toward conservation. The US president in recent history who made the strongest contribution to conservation was not a progressive, articulate, well-manicured, teleprompter-reading, charismatic president. The best environmental president was, in fact, Richard Nixon. Not because he was ideologically committed to the cause but because there was a very strong concern in the United States about clean air, clean water, endangered species, and so on. All the great laws that protect the environment were passed during the Nixon administration simply because he was pandering to a very concerned citizenship, not because he wanted or personally cared about the environment.

What Is True Conservation?

I have summarized the technical aspects regarding anaconda biology that may be used for management. However, I need to caution the readers about a common mistake of our day: Management is not conservation per se. A management program can be used for conservation, but a management program is not necessarily conservation unless it actually helps protect biodiversity beyond the sustainable use of the organism. The notion that any management plan is necessarily conservation has been promoted for the last few decades as part of the international economic agenda trying to capitalize on the increasing environmental awareness.[155] This has been especially true in the management of new commodities, such as wildlife. Management of traditional resources such as forestry and fisheries still goes by its original designation; they are still considered businesses, but those who manage new resources have often tried to appear more likable by presenting them as *conservation*, not as business.

However, there is plenty of evidence that wildlife management is not necessarily conservation. Consider a bird-watching operation located in the area where a very shy and rare species of bird nests. Birdwatchers flock to the site during the nesting season to see this rare species, bringing a great economic surge to the local economy. This could be thought of as a good example of conservation via the non-extractive use of the species. But let's consider that this rare bird is so shy that the constant observation by tourists compromises its nesting success; this population of birds would be literally "watched into extinction." It is profitable for the community, but is it conservation? For a program to be true conservation it has to have conservation as a goal and not just as a byproduct. If conservation is a byproduct, the system can easily stray into regular business that might not even be sustainable.

A wildlife management program can fall into any of three categories. First, a business can use an environmental commodity thoughtlessly until it runs out. Second, a business can uses some resource of the environment in a sustainable manner, but the goal is to make money, so it does not offer great economic incentives to the stewards of the land or ensure the long-term conservation of diversity. Third, programs can use the resources in a sustainable manner but also provide economic incentives such that the stewards of the land have good reasons, and resources, to protect the environment from other potential uses that are not sustainable. It comes down to the model of the operation. When the bulk of the economic incentive goes to the local communities, they will have both reasons and resources to prevent external enterprises from threatening the environment. What this does, in effect, is to vaccinate the system against unsustainable uses.

There is no doubt that the first form of management is not conservation at all. Lots of fisheries and forestry operations fall in this group; they simply plunder a resource until it is gone. Some biologists feel that both the second and third are legitimate ways to do conservation, while I feel that only the third one is true conservation. An operation that is sustainable and in compliance with conservation of biodiversity can, and should, take credit for being clean and sustainable, but just because something does not destroy the environment is not enough to construe it as a conservation program. What I am proposing is not an ideological stretch: *The goal of a conservation program must be conservation.* Economic gain could be a byproduct, or a means to do conservation, but it must not be the goal. This ideological issue has been, perhaps, willfully ignored in the conservation literature and conservation work.

Wildlife Management and Conservation: Let's Not Confuse Them

The problems of conservation and the use of wildlife are not detached from other economic and political issues, and we would be mistaken to try to address the former without considering the latter. Common tendencies are to use the natural resources for profit without a real environmental agenda by benefiting from the opportunities and even funds that conservation activities may have. Such operations are often not even sustainable but simply use the natural resources in a seemingly "green" manner. Those operations are even more harmful for conservation than other activities for two reasons. First, they use and deplete the natural resources just like others. Second, because they are done in the name of conservation, they create bad PR for conservation causes, drain funds from conservation activities, and distract attention from the real solutions of conservation problems.

In the next sections, I provide with two examples using reptilian representatives of wildlife to solve economic crises. First, I use the example of managing spectacled caimans in Venezuela,[148] in which I was involved for years. I also revisit the management of yellow anacondas in Argentina.[141] I discuss them briefly and frame them in the macroeconomic context where they belong, which allows us to understand further elements considered in the decision making. This is not intended as a comprehensive analysis of the programs; I'm just providing an overview of the critical points.

Spectacled Caimans in Venezuela

Since the mid-1980s the Venezuelan government has carried out a program for harvesting spectacled caimans. This program operated on private lands, where the owners hired a biologist to survey the population size, and, based on the population estimate (or other surveys of the area), Profauna gave a license for a given quota. The owner then hired people to harvest and process the animals. The skins were bought by tanners, who prepared the skin to *crosta* (one of the steps of the tanning process) and sold them to overseas companies that made the final product. This program provided some benefit to the landowners, the local workers who performed the harvest and transported supplies, the biologists who did the survey, and the tanners who processed and exported the skins. This program was based on a very prolific species that has a very high commercial value, is very easy to count and harvest, and belongs to a group that has proven to be fairly resilient.[148] In short, it is a *perfect* species for conservation management.

Regardless of the well-intended efforts of Profauna in running a biologically sound program, from the beginning Profauna was involved in a battle of wits with the poachers and other sectors that took advantage of the loopholes in the regulations. After the word got out that every square foot of caiman skin was worth $40 (remember, daily wages were about $3.50), there was no safe haven for the animals. Every improvement in the legislation was met immediately with new ways to circumvent the law. To illustrate this fact, I will relate one of the problems that the program had. Landowners sent their harvesters to trespass on other people's land to kill and market other people's caimans to keep their own populations high for future surveys. In response to this problem, Profauna decided to count the skulls and carcasses of the caimans that were harvested and match them with the number of skins as a way to ensure that the caimans were actually killed on the lands of the producer (carcasses are too heavy to carry on a burro's backs, which is the reason poachers only retrieve the skins they poached). This regulation immediately spawned a new breed of small entrepreneurs in the llanos. Their business consisted of carrying a truck loaded with rotting caiman carcasses that were then rented out to crooked landowners who had hunted caimans illegally and needed the carcasses to "wash" the skins they had poached. The government officials ended up counting the same carcasses over and over, matching them to different skins, on different ranches. This is only one example of the many tricks that Profauna had to uncover in their effort to implement the program.

Most of the people who were supposed to get involved in management and start protecting the resource for sustainability never perceived it as something other than an ephemeral source of wealth that was there to take advantage of while it lasted. Of course, this uncontrolled rate of harvest resulted in the population declining dramatically in many places (personal observations and reference 148). This decline, along with a drop in the international prices of the skin due to the success of alligator farming in Florida, brought the program to the brink of extinction, reinforcing the idea that caiman harvesting was indeed ephemeral!

Although there is little that the Venezuelan program could have done about the Florida farmers, a farming program cannot really compete with a ranching one. The investment needed to produce a foot of skin in a farm is far higher than what is needed to produce the same in a ranching system. Ranching can offer the products at far lower price than farming, even if the alligator skin is of higher quality than caimans. Flooding the market with inexpensive caiman skins could drive the alligator farms out of business. Economists call it "dumping." Once competitors are out of business, greater profit can be made by driving prices up. However, the ranching system was fraught with problems of management and the populations declined; that is what really brought the program to its knees. This program was unsuccessful not only because it failed to convince the locals that it was a long-term program and that by abiding by it they could obtain sustained revenues. It also failed in giving enough economic incentives to the local people to really protect the resource and to make the harvest of caimans something valuable to them. The tanners and land owners were the greatest beneficiaries, but the locals got only temporary and poorly paid employment.[156] Hence, there was no grassroots pressure to protect the resource and the program.

Not only didn't the program help the conservation of biodiversity, it also hurt the cause. It consumed a great deal of time, money, and effort, with not many results to show for it. Also, it resulted in a lot of bad PR for the conservation cause and conservationists. Associated with this program were open farming initiatives where the landowners built facilities to raise some neonates up to harvesting age. It worked OK for a few years, but when the skin prices came down due to alligator farms, landowners found themselves in a position where the taxes for selling every caiman were comparable to the price they were getting. Profauna, strapped for cash because the ranching program had collapsed, could not adjust its taxes to the new prices of the skin because it needed the money. The ranchers ended up turning loose all the animals they had and taking big losses for the facilities and for raising several thousand babies. No one in their right mind should try to talk any of these ranchers into ever again making any investment in wildlife management or conservation without jeopardizing his or her physical integrity!

The goal of the Venezuelan caiman project was not conservation but business. The government agency was under a political mandate to generate its own resources (see later in the chapter for more details). During the time the project lasted, Profauna met regularly with the scientific community seeking advice about management. For the longest time we had discussions about how the program was being misused by tanners and crooked landowners and all the things that had to be corrected. However, Profauna depended economically on the fund generated by the program. The tanners who regularly broke the law were the ones who supported Profauna. You don't really have the authority to regulate or to "put the smack down" on someone who is paying your salary. When Profauna eventually had to change the program radically because of the population collapses, it meant the end of Profauna. There is currently a program of caiman harvest in Venezuela, but only produces some 15,000 to 20,000 skins a year—a long way from the 100,000 to 120,000 it once produced. So, the original program, the facts show, was not sustainable. This proves that the goal of the program was not conservation but business. It could have been a "green" sustainable business if they had had the capacity to regulate it well, but because they did not, it ended up being an economic enterprise that actually diminished the resource it was using. It was, in fact, an example of the first case I was mentioning earlier: plundering as opposed to management or conservation.

Yellow Anacondas in Argentina

In 2002 a province in Argentina, Formosa, embarked on a novel program of management of yellow anacondas. The program is based in a large swamp with very little access, inhabited by several indigenous and rural communities. The way it was set up, hunters were allowed to hunt any animals above a certain size. There was no limit on the number of animals per hunter or per season.

When the program started, I felt that it was more of an extermination program than one of sustainable harvest. It harvests larger animals, mostly females during breeding year, with no limits or quotas or number of captures per year. However, time has proven me wrong. The program has continued for more than 15 years, with no evidence of decline in the population.

Apparently, because the swamp is so large and the access is so limited, they estimate that if any local depletion were to occur, it will quickly be fixed by immigration from the vast area that is out of reach for the hunters. It is some sort of a sink/source system that seems to work well. So, even though it may be unorthodox, by all metrics this program is sustainable. I stand corrected— and happily, I may add, because the program delivers a lot of good for a lot of people.

The program has produced substantial economic benefit in a region that has very little economic activities. Because of the abject poverty of the area, the low capture rate and the low prices of the skin do not hinder the interest of the locals in participating in the program. This program has produced a lot of good for the local community. Anaconda hunting can bring up to $75 a year for a family that lives on a yearly budget of $150, a 50% increase in their economic well-being! Also, law enforcement officials reported that the rate of cattle robbery, and common crimes, had dropped to historic levels since the program began because the local people had legal ways to make some money.[139]

As good as these consequences are, they are no different than what would result from any business moving into town. Let's say that a meatpacking industry moved its operation there. It would decrease poverty, create jobs, and relieve pressure on unsustainable uses of the environment. When these goals are accomplished by a sustainable use of a given resource, it is no different than when they are accomplished by a traditional commercial enterprise. They are both good for the people, and if the commercial enterprise does not put undue pressure on the environment or the exploitation of the resource, it is sustainable and good for the environment because people with jobs don't have to hunt unsustainably to obtain meat. There is nothing wrong with this as a plan for development and generating revenues for the people.

Now, let's imagine that some corporation wanted to set dikes to dry out large areas of the swamp to, say, plant oil palm or sugar cane for the production of biofuels. This operation will destroy the habitat and eliminate the population of anacondas as well as decrease substantially the biodiversity of the region. However, it would also offer permanent and more reliable employment with superior income to what the locals make with the anaconda extraction. Will the incentive offered by the anaconda program, as it exists now, be enough for the local people to oppose the corporation from taking over and destroying the habitat? Will the local people have the reasons *and the resources* they need to succeed in protecting the environment? Can they fight the takeover of the land in court? Will their current income be enough for them to overcome the offering of more, and more reliable, easier employment? The answer to these questions will tell us whether this program is a conservation program or just an economic enterprise that uses a natural resource, sustainable as it may be.

When the program began, locals were paid $4 to $5.40 per skin, depending on its size. This is great for an economy that is depressed and for people who have no other source of income. Middlemen made $50 per meter of snake. Thus, a 3-meter-long (9.8-foot-long) snake produced $4 for the hunter and $150 for the middleman.[2] Eventually these skins were turned into purses and other commodities worth large sums in the international market (as in $2,000 a purse). So, the structure of compensation was not that different from that of the spectacled caiman project in Venezuela, with most of the profit going to tanners and foreign fashion companies and very little put into the local economies. It is clear who the great beneficiaries of this activity are and what economic sector was in mind when the program was set up. A program aimed at providing resources for the stewards of the lands as a conservation strategy should maximize their benefit and income. The problem with a system that is set up to help the economic elites is that they can very easily move their operations to other places and are not committed to the maintenance of that piece of land. Even local businessmen can easily invest in other operations (e.g., stock in the palm oil plantation) and still do not fight to protect the environment. If we really want to protect the environment, we need to make sure that those who belong to the land and whose livelihoods are linked to the health of the ecosystem make enough money to have both the reason and the resources to protect the environment. So, I am saying that a true conservation program must be one that inoculates the system against external, unsustainable, takeover.

I am not opposed to some economic activity that may use the environment in a sustainable manner, but it should not be presented as conservation. What I need to draw attention to is the fact that often these economic activities may receive conservation funds or other benefits for being considered a "conservation program." This drains the ever-smaller resource pool for conservation into other activities, erodes the trust in conservation activities to protect diversity when these end up failing, and distracts from work on real solutions.

Intag: Sustainable Agriculture in Ecuador

An example of true conservation is a community in the Ecuadorian Andes where farmers have established a community-based operation in the Intag region. The area is located in pristine cloud forest with mega-diversity of plants and all sorts of animals native to the cloud forest. The community has set up an operation that produces noncertified organic coffee, produce, sustainable ecotourism, and hand-crafted art based on local resources. Their produce and coffee are not "certified organic" because obtaining the organic certification would cost a whopping $20,000 fee, not counting the expenses of the person who does the certification. However, their produce and coffee are not only organically grown but are also grown in the forest without cutting it down, and working with the ecological processes present in the forest. Shade-grown coffee takes longer to grow and the production per plant is not as high as if they cut down the forest, but they are getting some profit from the forest while maintaining its integrity and its natural processes. So, it is a step ahead of regular organic in the sense that no chemicals are used, organic, but also the plants are also grown in harmony with the existing ecosystem. They have not made great fortunes with it, but they do live in the forest using agricultural practices compatible with the forest, and the very forest and their sustainable operation are part of the product they sell in ecotourism.

This land holds significant copper ore, but the local people did not want the ore to be exploited because it would destroy the environment and their livelihood. Around 2004 Ascendant Copper, a Canadian company, started trying to remove the people to extract the copper. Unfortunately, the gadget revolution that has given us so many nice toys like cell phones, tablets, and what have you had put tremendous pressure in the world supply of copper. By now, there is little ore left and the few deposits that are still around are in pristine habitats. Copper mining is a truly dirty business that produces all kinds of pollution as well as sediments and toxic waste released in the environment. If Ascendant had its way the livelihood of these people would suffer terribly, and the forest where they live would be changed forever. However, the farming operation gave the modest community in Intag a livelihood good enough that people were not tempted. The copper company has offered money and jobs to the members of the community, but they have stuck to their position and rejected the offers. True to form, Ascendant hired thugs to beat up and harass community leaders. Some of them had to escape into the forest fearing for their lives, but the community still held on to its position. Following the established playbook, the company enlisted the government and declared that the community leaders were thugs, delinquent of various kinds, and people who needed to be "brought to justice." The leaders hid in the vast forest and responded with the resources they had in the legal system and the community and did not budge. In 2007 the new government in Ecuador, a government that proposes a different development model than the usual, banned Ascendant from doing any business in the country. Other copper companies have tried to start operations in the area, but the local community has had both the means and the reasons to fight them off and protect both their livelihood and the environment. So long as there is the need for copper, Intag will always be endangered, but so far as the people have their community operation they can fight to defend it.

What would have happened if the operation in Intag had been led by a large corporation that managed the agriculture and the tourism? It is unlikely that a large corporation would have been involved in the community's environmentally friendly operation because it produces less profit than a larger-scale one, but let's assume that they did. Let's call it the "Sustainable Company." It's more than likely that when Ascendant came along, Ascendant would have paid enough money for the Sustainable Company to sell Ascendant its operation. Because the Sustainable Company is not tied to that piece of land it can, with enough money, move to another place and start another sustainable business elsewhere. The local people who worked for the Sustainable Company would have been laid off, without the means to make a living or resources to oppose the actions of Ascendant. Eventually they would have to work for the very company that was destroying their land. What the community-based operation of Intag did was that it effectively inoculated the area against hostile takeover because the profit from the operation stayed in the local community. That is true conservation.

The Status Quo Conservation Practices

Conservation programs whose only goal is to conserve biodiversity do not represent the standard practice in mainstream conservation work. In fact, the Argentinean program falls squarely within a good conservation program under current standards. In mainstream conservation work, any activity that uses the resource sustainably is normally called conservation. What I am proposing is raising the bar of what we call conservation. My using the Argentinean program as an example is not intended to discredit a worthy enterprise that has rendered good results for the local economy, and the people, without harming the environment. It is to bring an example from the anaconda world to show what I consider to be a weakness of the conservation approach we practice worldwide.

As mentioned, in the past four decades conservation work, funding, and conservation programs have increased all over the world at an accelerated pace. Yet, the conservation crisis only deepens and deepens despite all these efforts. Clearly, we need a different approach. I am proposing that we need to change what we are doing and start practicing a radical approach to conservation. We need to focus on the roots of the problem and address it where it is.

The Problem of Conservation

As identified earlier in the chapter, the main conservation problem in Latin America can be traced back to poverty. Going to the root of the problem, the following sections address the economy head on as it is an intrinsic part of the problem of conservation. Most academic biologists as well as the lay audience consider economics and politics as bad words: because we are not trained in their use and implications, we prefer to avoid any issue that can lead to their discussion. For this reason, when we discuss the economics of conservation, we usually talk about economic incentives for conservation and sustainable use of the natural resources but often do not go any further. We identify the cost of the commodity involving nature, what it represents for the environment, and what economic incentive the local people receive, but we often fail to place it in the bigger macroeconomic framework where it belongs. This is the reason that the solutions we offer often fall short of meeting the real economic needs of communities and thus turn out to be ineffective, or at least vulnerable to other pressures.

Macroeconomics for Dummies

For the last 60 years, international economic agencies (IEAs) such as the World Bank, the International Monetary Fund (IMF), and the US Agency for International Development, to mention a few, have been sponsoring development and giving grants or loans for developing countries to increase and *aid* their economies. The idea is that with the money injected into their economies, the developing countries will be able to create industries, factories, and other source of employment that alleviate the poverty of the area. Once the economy has been activated, the countries can pay back the money. The market will take care of everything: once the country starts doing some business with the aid received, there will be jobs and cash flow and the poverty will go away so the countries can repay their debts.[157-159] Once they are solvent, conservation will follow by itself, just as has happened in industrialized countries. The notion is that wealthy countries don't need to rely on the unsustainable use of their resources because they have industries and what have you that provide income for everybody.

However, the results have not been quite as expected after 60 years. Instead, the countries that have received more economic help have experienced a dramatic increase in poverty. Countries that abide by the macroeconomic model fall deeper in debt and often spend all their gross product paying off the debt without solving their problems, or repaying the principal. Many countries identified as "failed economies" are simply victims of this system.[160-163] How can this happen in countries that are getting so much help? The truth is that this *help* does not come without some strings attached. Often the loans are conditional, with the countries giving up some of their sovereignty in decision making. IEAs often request that countries adopt a number of internal economic measures. Common measures are decreasing or eliminating internal subsidies to their agriculture and goods, dropping trade barriers, allowing international companies to operate in the country freely with little or no taxation, and instituting deregulation. They require the privatization of the school system, the health care system, utilities operations, ports, and other infrastructure. This loan requires exceptions to be made in environmental regulations for businesses, elimination of social benefits such as social security, and relaxation of labor laws, to mention a few. These measures are often called structural adjustment programs (SAPs), or more recently austerity measures, and are imposed on the people when the government accepts the loan, credit, or other kind of economic "aid."[159,162] So, the people in these countries actually experience a lower standard of living since SAPs eliminate the social safety net and other governmental programs that kept their heads above water.

In exchange for accepting the SAP, or the austerity measures, the IEAs give loans and credit to the companies that will bring jobs and capital. The rationale is that the money from the companies will activate the economy. However, the countries do not receive the money to invest at their own sovereign will; there are often limits placed on how it can be used. Often, the money must be used in hiring international companies, in no-bid contracts, to do large development projects (roads, dams, etc.). These are often companies associated with the government of a developed country whose job is to secure sales for the country of origin.[164] So, a good part of the money (about 40%) is given without bidding to a predetermined company and never reaches the country it is supposed to benefit. The money is not injected into the local economy, where it would have produced local economic activation. Instead, it is just moved from one bank account to another in some developed country that does not have to pay the loan or the interest.[159,164,165] Since only a fraction of the money reaches the country that gets the "help," but the country needs to pay the totality of the money plus interest, there is only a net flow of capital out of the country, not in. So debt acquired by the government to hire private companies is now public debt and the taxpayers are responsible for it.[164] The company that administers the loan hires employees at minimum wage, but since the SAPs lower the wages, social assistance, and workers' benefits, the minimum wage does not really solve the problem of poverty in the area. People have jobs, but the prices of regular commodities (produce, water, housing) have gone up because they are now in a globalized market, so their standard of living is even lower than it used to be.[159,166–168] Clearly, poorer people will resort to unsustainable uses of the natural resources if that is the only thing they have left.

Because the SAPs open the country to the international market, the national companies are bought by international mega-corporations since the national companies cannot compete with the foreign investors once there are no trade barriers. The small and local companies then must face competition with multibillion-dollar transnationals that leaves them bankrupt.[169] Imagine a small phone company in, say, Argentina, competing with AT&T. The Argentinean company has no chance of succeeding, and the owner will be forced to sell the company for little money and will become, at best, manager of the business that he or she once owned.

Increasing poverty linked to this macroeconomic agenda may lead to environmental degradation by forcing people to use the resources in an unsustainable manner. However, this is not the only way in which these trade policies hurt the environment. The companies now have liberty to dump their waste in the watersheds, due to the imposed lowering of environmental standards included in the SAPs,[164,170] and their activities harm the habitat where the local people live, bringing problems of disease and pollution that affect the whole ecosystem and kill the local wildlife.[163,166,168,171] In general, the new international companies lower the quality of life of local people and jeopardize their chances of returning to their former lifestyle. Many of these companies are temporary, like mining or logging. When the company leaves the country, it leaves behind pollution, destroyed habitat, unemployment, and even more poverty than there was to begin with.[164,172] The effect of these economic aids by IEAs has been compared to the use of anabolic steroids in sports. It can produce a temporary spike in performance but is bound to produce problems and results that are detrimental in the long term.[173] The impoverished nation will then return to exploiting whatever is left of the environment in an even less thoughtful manner.

These economic strategies represent both sides of the coin. The side that faces the developed countries is called globalization, while the one that faces the developing countries is called neoliberalism (which is better known as neoconservatism in the United States). Regardless of the name used, it is strongly linked to extreme poverty in developing countries, as shown by the evidence from all the countries that have abided by it.[157,160,174] A good example of this sad situation is Argentina, which embraced this agenda wholeheartedly during the late 1990s and even became a model for all developing countries due to its wealth. In 2002, however, the bubble burst in the Argentinean economy, leaving the country in great poverty and great economic turmoil.[169] Later on Greece, Spain, and Portugal faced similar problems.

The crash of the Argentinean economy was followed by political upheaval and more economic and social turmoil. The influence of IEAs producing extreme poverty through the application of SAPs can be seen in developing countries throughout the world, often leading to similar social unrest and political problems. This has been the cause of the recent popular uprisings in South America that ended up toppling presidents in the last decade in Argentina (2002), Bolivia (2003, 2005), and Ecuador (1997, 2000, 2005), just to mention a few. Needless to say, during times of economic turmoil and political upheaval, conservation takes a back seat.

Neoliberalism and Conservation in Latin America

Economy and the Environment

Why is all this political and economic brouhaha important for conservation? Well, when countries have these kinds of problems, it negatively affects conservation in different ways. Poverty leads people to the unsustainable use of natural resources. When hunger strikes, no amount of environmental education or enforcement that can protect the environment. People will resort to the unsustainable use of the environment as a first resource.[136,168] Also, during times of political upheaval, countries tend to set aside conservation programs, environmental education campaigns, or environmental policies and enforcement. All of this produces negative effects on biodiversity.

Right about now is the time when people who are uncomfortable talking about this issue tell me that I need to shut up about economics and focus on anacondas because that is what I really know. I would reply that the economic upheavals that ruined families and bankrupted nations and whole economic zones was not done by herpetologists who meddled in what they did not know. It has all been done by the best and brightest economists and politicians in the world! This can mean one of two things: either they don't know what they are talking about or they know the consequences of their actions but do not care about the suffering they created. So, by caring about the environment and the livelihood of the people, I am automatically more qualified or better suited to comment on the matter. Plus, as I said earlier, economics and politics are not something I sought out but something that happened to me as I was in situations that left a profound mark on my life. I did not learn issues of macroeconomics and politics in textbooks but on the streets, experiencing the consequences in my flesh. That is more than most scholarly economists, sociologists, or political scientists in the United States can say.

A Social Explosion

As indicated earlier, I have experienced five *coups d'état* and one social explosion. I will never forget the speech I saw on TV on February 2, 1989. Carlos Andres Perez had been elected Venezuela's president for a second time. He had previously been president from 1974 to 1979. It was a big deal because it brought memories of good times and economic wealth when he was in power the first time. All Latin American presidents were invited to something that people dubbed "La Coronación" (the crowning). The speech was from Fidel Castro, who was Cuba's president at the time. It struck me for the unusual content. He said (paraphrasing): "Latin America does not have to worry about a nuclear holocaust. It will not be a nuclear bomb that will kill people in Latin America. Hunger will. Hunger is the enemy of our countries and it is what we need to be worried about." This was most unusual to me because in 1989, we all were worried about the nuclear holocaust. The Cold War was still going on and we all feared that there could be a nuclear war that would have unpredictable consequences. Fidel continued (paraphrasing): "We need to fear a social explosion and the consequences that it may have on our people." That was the first time I heard that expression, and little did I know how close I was to knowing exactly what it meant.

On February 27, 1989, 25 days later, Carlos Andrés passed his first economic measure, accepting the SAPs imposed by the international lenders. It was a neoliberal package that he had accepted in exchange of international "aid." Among other austerity measures, it included a raise in gasoline prices. Since the oil company is owned by the country, gas prices were held at low prices, not market prices, in order to keep the economy running since everything relies so much on energy prices. This increase in gas price resulted in higher costs for transportation that people needed to get to work and triggered a chain reaction on everything else. People arrived at the bus station only to learn that the price of their bus ticket had tripled overnight. This was a heavy blow to the working classes, the majority of whom lived paycheck to paycheck. This increase in their bus fare would have meant that it would cost them more to go to work than to stay home, when you take into account the cost of buying lunch. At first the people refused to get on the bus and waited for the next one, but the next one also had higher fares. As the time passed people continue to gather at the bus stations, their anger rising as they saw the unfairness of the situation and worried that being late to work could cost them their job. Yet that job would not make ends meet with the increased transportation cost.

Once you have enough angry people in the streets, all it takes is for one of them to throw a rock through a window display of a store for all hell to break loose. That happened. In no time the whole city was ablaze with people looting stores and ransacking everything in their path. The police tried to contain them at first, but seeing that they could not, some police joined in the looting, as they were poor as well. The mass of the poor and oppressed had seen an opportunity to fight back and were taking it. The media, always so helpful, broadcast all over the country what was happening in Caracas, and in minutes the whole country was on fire, with people in the streets rioting and looting stores and businesses in all major cities in the country.

With the police overwhelmed, Carlos Andres, following the handbook, ordered the National Guard to repress the people with full military force. The masses did not have weapons. This was not a revolution of any sort; it was just angry people in the streets. Unarmed as the people were, they had enough anger to make up for their lack of guns. The Caribe blood running in the Venezuelan veins rose up and the people fought back. A Dante-esque situation ensued, with the angry people throwing rocks and the National Guard, armed with high-tech weapons, shooting live ammunition into the angry, unarmed crowd. Rocks are no match for rifles. Order was restored in most of the city after the first 48 hours, but at the cost of countless lives. The people returned to the poor neighborhoods and shantytowns, where small streets and crowded buildings hampered the military forces. In some cases, the people responded to the gunfire with their own weapons, if they had handguns in their homes, or Molotov cocktails they prepared once they were cornered in their neighborhoods. Caracas was split into several areas of struggle in pretty much all the poorer neighborhoods, where the discontent ran deeper. It took another 3 or 4 days for the National Guard to prevail and for "peace" to be restored. In the first day, the morgues overflowed with bodies, and after the second day the morgues were not even taking them there anymore. It was not uncommon to see corpses floating in the Guaire River that bisects the city from west to east. It's still not known how many people were murdered to restore order on that fateful day.

The day of the social explosion, I was just back from the field and joined some of my friends to chat and swap stories about our field projects. We met for lunch, unaware of the situation that was breaking out in different parts of the city. Before we were done with our meal the owner of the establishment asked us to leave: he wanted to close because he was worried about what was happening. Then we realized what was taking place out in the streets. I drove my friend Cesar Molina to his house in La Vega, one of the poor neighborhoods on the west end of the city. There was hardly any traffic, but when we got to the entrance of La Vega we found a group of military guards who were seizing the area. There were barricades preventing police or military vehicles from coming in, and the people had established positions on top of buildings and houses surrounding the area. The guards did not dare to enter. Not wanting to be anywhere near there, I drove against the traffic direction in a street that came out of La Vega. Some 50 meters in, I ran into another barricade, but the people manning the barricade recognized Cesar and opened it up for me. I drove deep inside La Vega, far from the fighting front.

When I parked in front of Cesar's house, I had two flat tires from driving over the broken glass they had put in the barricade, and I was stuck in one of the most dangerous neighborhoods of Caracas, on the most dangerous day of the century. Yet, inside La Vega, the mood was one of tranquility and even joviality. There was an alleyway that led in and out of La Vega away from the fighting front, and people were going down to the nearest mall and bringing back all the looted merchandise they could carry. Some shared their loot with friends or swapped goods as they needed. I saw a true feat of strength from a fellow who could not have weighed more than 50 kilograms (110 lbs) if he were soaking wet: He was hauling on his back half a heifer he had gotten from a butcher shop. As things felt so homey, my curiosity kicked in and I wanted to go down to the area and explore what was happening. The words of Fidel Castro came back to me in full force: "We have to fear a social explosion." That is what the old fox meant! In science, if a variable, or set of variables, can be predicted by other variables, then this subject is susceptible to straight-forward scientific inquiry. Knowing that the situation could be predicted, it meant that there was a set of independent variables that had produced the event. Fidel clearly had read the writing on the wall ahead of time and was warning the world about it.

As I followed the crowd down to the mall, there were people carrying mer-chandise in the opposite direction as well as other people returning to the mall for more. I walked into a warehouse where people were still taking as-sorted merchandise from the shelves. The guards had tried to protect the mall at some point from the looters but had given up and conceded this ground to the people. The air was thick with the smell of tear gas, yet the people seemed to be immune to it, walking at ease in and out of the areas where I could barely breathe. I turned the night into an event of scientific observation to make the best of it. I observed the optimal foraging practices of the people in the stores and how they chose the best items, maximizing the efficiency of their car-rying load. I saw social collaboration in how the people shared the weight of heavy items and split them up later, and how younger, stronger people self-lessly helped older, weaker individuals to carry heavy items. About 4 in the morning, I was starving and asked a stranger carrying a sack of Cheetos for a handout. He tossed me a couple of bags, which I wolfed down in an instant.

Whenever there is a situation of looting, even in the United States, it is not uncommon to hear the expression "Shoot to kill looters." Harsh as it sounds, most people accept it as just something that happens in the case of looting. Yet, when I hear that expression, what I really hear is "Death penalty for shoplifting, and without a trial." It was obvious to me that the looters were not a crazed crowd of non-human beings that needed to be eliminated; they were just people making the best of the situation they had been presented with. Later, I learned concepts such as "structural violence" that allowed me to place what I had learned within a formal cognitive context, but I had seen it firsthand and knew exactly what it was. So, while my transcript does not show that I took advanced courses on globalization, macroeconomics, and free trade, and I have never read a textbook on the matter—boy, do I know what they are!

Neoliberal Agenda in Venezuela

To better understand these issues, I will bring in the larger worldwide context for the conservation programs we have discussed. The decline of caiman populations was evident from the first few years of the program;[153] however, Profauna chose to ignore repeated warnings from the scientific community that the caimans were being overexploited. The likely reason for this is that Profauna depended economically on the revenues that caiman skins were producing. If the project were halted to let the populations recover, the administrators of Profauna would have pretty much eliminated their own jobs since their salaries depended on the revenues from the skins. There is something basically wrong when the people who are to decide on the administration of a resource are depending economically on the decisions they make. How can they be expected to make objective decisions for the environment when their own interests and paychecks are directly linked to those decisions?

This is more than just a simple problem in the way the system was set up in Venezuela; this is an example of a larger structural problem with the macroeconomic approach. In the late 1980s Venezuela was under a strong neoliberal grip.[175] The recommended measures for economic development (SAPs) demanded that the government did not sponsor research or any other activities that were not linked to administration of the resources; *self-financing* was a very common word among all the governmental institutions. If the caimans were worth money, there should be economic investment to do all the research and finance all the needs of the program. So, the government did not have to finance anything (including the salaries of their personnel). The law of supply and demand should be the only ruler of the system. The neoliberal agenda contends that the market is an invisible hand that solves and takes care of all the problems.[160] However, the way the system was set up, the managers had to compromise their own salary if they took any action to stop the program. This was the main reason that the program was not changed when the decline in caimans became evident. In fact, when the program was finally cut dramatically due to the low density of caimans and changes in the international market, Profauna underwent major structural changes and major downsizing and eventually disappeared.

We find a similar problem in the program to harvest yellow anacondas in Argentina. During the 1990s Argentina had embraced the neoconservative agenda, giving up all its assets and agencies, dropping all taxation barriers for international corporations, and benefiting from the temporary economic boost it gave to the economy. Toward the end of the 1990s, however, the economy was slowing down, and the country was facing serious economic problems. At the end of 2001, the IMF pulled out its mission in Argentina, and the country crashed into the worst economic crises that anybody in the country could remember. Having no assets left, the people, the provinces, and the country needed income, money, and investments to solve the crisis.[169]

The province of Formosa is the poorest economically and is also the northernmost province of the country, so it is the closest to tropical latitudes. It has the greatest wealth in biodiversity. When an offer to exploit its resources came along, the residents could not really afford to turn it down. The associations of tanners in the country created a fund of $100,000 to harvest the *curiyus* (yellow anacondas) in Formosa. This was a juicy contribution to a very stagnant economy; the province was to use the money to hire the people to do the study and administer the harvest. The neoliberal constitution (1994) at the time established autonomy for the provinces over their own natural resources, taking the decision making away from the central government of the country (Article 124)—the good old "divide and conquer" strategy. This is the heart of the problem. Local authorities might be persuaded to accept offers that might not be in the best interests of the country or the environment.

We cannot even blame the government of the province of Formosa. The neoliberal constitution brings the tragedy of the commons to a new level. If the province of Formosa did not accept the deal, there were other provinces—Chaco, Misiones, or Corrientes—that could have benefited from it. The priority of the whole system is the urgent need to produce cash to ameliorate the economic crisis; politicians cannot afford to take a long-term perspective. Turning down such a business opportunity was not an option. The level at which countries protect their environment and natural assets is tightly linked to their economic wealth.[176] The SAPs imposed by IEAs for more than a decade had bankrupted Argentina to such a level that any program that would produce some money had to be accepted. To their credit, the organizers of the program managed to put together a program that provides for the needs of the people and produces some funds for the local economy while not depleting the resource. In the neoliberal dictatorship the country was in, one could expect no more.

On the other hand, the government in Ecuador was following the neoliberal agenda to the letter in the early 2000s. At the peak of the problem in Intag, the National Guard itself came to the community to harass the leaders, doing the bidding of the corporation. It was not until 2007, when a new government was established that openly rejected neoliberal policies, that Ascendant Copper was expelled from the country and the community members were allowed to continue their operation without fearing for their lives. Even if pressure to use the copper ore remains, and will remain, so long as there is such a worldwide demand.

The Dark Side of Macroeconomics

When we talk about the conservation problems in Latin America, there is often a point at which we mention poverty and corrupt politicians. Then we throw our hands up, as if poverty and political corruption were unrelated. That is not quite the case. In Latin America for the past six decades there has been natural selection benefiting corrupt politicians. No, really! I am not even talking about the higher fitness of corrupt people who make more money. I am talking about a low survival of noncorrupt politicians. In the past 60 years, whenever there has been a Latin American president willing to stand up for the best of his country and not give in to the pressures from the corporate elites, he always ends up dying in a freak plane crash, being assassinated (or at least toppled) by some totalitarian regime, or being removed via suspicious elections,[165,177] or is "asked to resign" after manufactured civil unrest, some externally sponsored "color revolution," that ends up toppling the legitimate president.[178] If any of them should return to power, they had learned their lesson.[179] The totalitarian regimes that followed always have great relationships with, and have all the support of, IEAs despite their terrible record in human rights and their nondemocratic origin. They promote more of the corrupt policies that produce poverty and that lead to more environmental degradation. So, the combination of corrupt politicians, poverty, political unrest, and misery that we often blame for environmental neglect has everything to do with geopolitics and the macroeconomic international agendas in developing countries. They often show a very different face to the developed world than to the developing one.

But What Can the People Do?

Often, people feel frustrated at this point of the conversation. We like biology and wildlife, but politics, assassinations, and *coups d'état* are way beyond what biologists and people who love nature want to read about. However, whether we like it or not, they are present all over Latin America. Whether we like it or not, they are extremely relevant to the main problem of producing and maintaining extreme poverty and its mandatory ugly offspring: environmental degradation. The truth is that there are many unpleasant things about working in conservation, from having to be involved in killing or culling of organisms we may revere, to seeing species go extinct, or having to compromise on protection of some habitats and seeing others being destroyed by human activities. These issues of politics and macroeconomics are simply another set of ugly things that we need to endure if we really want to talk about conservation. If we think about it, ecology as a science does a lot of looking for trends and relationships among different variables; economics and dirty politics are just other variables to throw into the equation.

Conservation Fallacies: Getting to the Moon

Choosing to ignore or deny the impact of macroeconomics in conservation will not help us understand or contribute to a solution. We cannot really understand the problems of conservation unless we address the root. We might feel inclined to ignore the big problems and try to address the ones that are easier, cheaper, or simpler to solve, the ones that are in our reach. But that is the fallacy of Tylenol conservation. Taking a painkiller is cheaper and less scary than going to the dentist, but it does not solve the problem. In fact, taking a painkiller guarantees that the person will have to take more painkillers later because the problem will get worse. This is a very common trend among biologists when we attack big conservation problems. We may choose to apply the simpler, cheaper solution that is at hand and that seems like a help for the conservation problem. You may hear a conservationist saying, "We cannot afford to stop using fossil fuel, but we can use more efficient engines." Notice the presence of both the problem of affordability and trying to compromise the problem without offering a solution that addresses the root of the problem. These are the trademarks of Tylenol conservation. Yet, most conservationists out there will sign off on these sentences I just wrote!

The problem with this approach to conservation is not only that it doesn't work; it's that it distracts us from looking for solutions that really work and it ends up being worse than nothing. At the risk of over-methaphoring the reader to death, I will provide another one that I often use in my classes to illustrate how harmful some well-intended actions can be if they do not solve the problem. I tell the students that we need to get to the moon for a very good reason. Then I stand on a chair on the classroom and ask everyone for praise since I have gotten a little closer to the moon. I often push the theatrics a bit more by putting the chair on my desk and climbing on top of it, getting even closer to the moon and asking for further praise for my accomplishment. We may even start a movement, getting everybody to stack the chairs and use ladders and ropes to stabilize the stack. We can do a fundraiser to buy more chairs and more ropes— and every time we are getting a little bit closer to the moon! Sooner or later someone points out that I am still very far from the moon. I then mock-chastise the student: "Well, you are so negative! This is better than nothing. What are *you* doing to get to the moon? I have gone a lot farther than you. Instead of attacking me for partial success, you should be joining me."

Eventually we discuss it. By stacking chairs on top of each other, I am getting closer to the moon, but I am not getting any closer *to get to* the moon. In fact, the first thing one needs to do to actually get to the moon is get off those silly chairs and get oneself inside a rocket that can actually get a person to the moon. Anything other than that is not helping and is actually worse than nothing. All the money and time spend in stacking chairs are wasted and would have been better used to get me to the rocket.

Real-world examples of this error we make in conservation are the widespread tendencies of scientists to identify areas of high diversity that need to be protected for conservation and try to pass legislation to protect them, when the real solution is not protecting the land but removing the pressures that threaten the land. Identifying the need for protection or protecting the land will only serve, like a painkiller, to delay the problem; it is not a real solution so long as the pressure on the land, the real conservation problem, persists. If there is any doubt, let's consider the case of the Arctic National Wildlife Refuge. At the end of the 1990s, after a lot of research documenting its uniqueness and years of intense lobbying by environmental activists, it finally acquired legal protected status. At the end of the Clinton administration it was protected "in perpetuity." This was considered a great victory for the environmental movement and an example of "how things should be done." A couple of years later the legislation was abrogated and now it is, again, available for exploitation, and we are back where we began. The campaign to protect it "in perpetuity" lasted longer than it lasted under protection! I am not saying that the protection of the Alaskan wilderness was not a great and well-deserved triumph of the environmental movement, but, as the facts have shown, it did not solve the problem so long as the root of the problem (dependence on fossil fuel) was not addressed.

These measures of conservation that only produce the illusion of a solution can be found on the "best" cutting-edge conservation planning around the world. Consider the looming problem of global warming. Conservation biologists and conservation advocates wasted 20 years pushing hard to get the US government to subscribe to the Kyoto agreement, yet all studies and scientists agreed that even if the world were to abide by the Kyoto protocol we still would be way short of the mark we need to stop global warming. It falls short of what we need and, in the meantime, it drains all the resources and efforts that the conservation movement needs to move toward a real solution. World authorities at the service of the energy industry skillfully produce new proposals (i.e., the Paris accord) and new international panels where everyone's attention is wasted. While people are distracted in this halfway solution, they don't work on the real business of solving the problem: stop consumption of energy. This is not unlike a street magician who makes a coin disappear and then pulls it out of your ear!

We have the same problem in the tropics whether we are fighting gold mining, poaching, unsustainable logging, or any of the other environmental threats. The problem is not any one I just mentioned. The main problem throughout the tropics is none other than the extreme poverty of the people in the area that makes them very easily tempted by these activities. We may obtain partial victories in protecting a piece of land, with education, demonstrating that a given mining company is not good, or providing some economic relief for the locals through wildlife management or tourism. However, those victories provide only temporary relief for the crisis and are bound to fail eventually if we do not find a true and lasting solution for the poverty problems that consume the region. Despite the great efforts the community has made, Intag continues being threatened by other copper companies. Hopefully their practices are not as unscrupulous as those of Ascendant, but the problem has not been solved.

Conservationists around the world frolic furiously with all kinds of Tylenol solutions to the problems such as ethanol, biodiesel, compact halogens, and what have you, but very few are really bringing the point home of the real solution: cut energy consumption by changing our lifestyles and redefining our society to meet the energy limitations of the planet. Someone could say that this is too difficult, it's unachievable or unaffordable, and so we need to settle for less—just like someone with a toothache would say that a root canal is too painful or too expensive, so a painkiller is more affordable. But can he really afford *not* to have the root canal?

The Question of Ideology

Few people will argue against the notion that at the core of most (all?) conservation problems we can find one or many of the following: overconsumption, poverty, corporate greed, pollution, corruption, hunger, and unfair distribution of resources. Few people would disagree that all of these are capitalism's despicable close relatives. Brian Czech has presented theoretical arguments and data demonstrating that an unlimited growth economy is not compatible with conservation.[180] Yet, very few people dare to propose a system different than capitalism and a free market economy, even though it can easily be shown to be responsible (at least partially) for the great majority of our conservation problems. All biologists will agree that consumerism is at the heart of most of our conservation problems. All economists agree that consumerism is the beating heart of capitalism. If you stop consumerism, capitalism will stumble and fall. So, people who know about conservation agree that consumerism is the problem, and the people who know about economics tell you that consumerism is needed for capitalism. If we put them together, it is clear that success in the conservation struggle requires getting rid of, or at least substantially changing, capitalism. Anything short of it is not addressing the problem and is, thus, Tylenol conservation.

Looking at Data

In case the reader still needs convincing, I will bring up a dataset that should eliminate any doubt unless the person is truly blindfolded by ideology. Figure 10.7 shows carbon dioxide (CO_2) emissions in the past few years.[181] We can

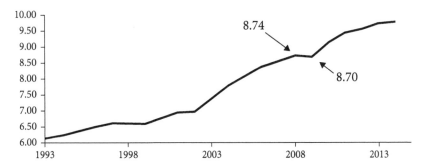

Figure 10.7 Emissions of CO_2 in billions of tons of carbon per year in the past few years. Notice the dip after the economic collapse of 2008.[182]

see the familiar trend of a steady increase in emissions over the past 15 years or so. The chart covers the time span in which great conservation activities took place: international promises to cut down emissions, multinational agreements to use less energy and reduce how much we dump in the environment. In this timeframe, there have been lots of technological improvements in the use of solar energy, more efficient batteries, more and better electric and hybrid cars. The number of hybrids and more efficient cars has increased tremendously during this timeframe. In the past few years there have been ground-level organizations to cut down emissions. International social media have driven events like "a day to turn off the lights" and "skip a day driving to save energy," and there has been a substantial increase in the conservation awareness of large sectors of the population. Yet none of these have produced any noticeable decrease in the amount of emissions we produce! The only notch in the trend occurred at the end of 2008—specifically, on September 13, 2008, when the world woke up to the news of substantial banking fraud among major players. The world economy teetered for a few days, and for a few days we did not know if it was the end of capitalism. For as long as 10 days we did not know what was going to happen next. Investors became jittery, and the production of some goods stopped. A short-lived hiccup in the economy was enough to produce a measurable effect on the amount of emissions we create. Nothing else has produced anything remotely similar to that. These are data, not ideological rhetoric. What the data show is that none of our efforts to stop global warming had any measurable effect on the amount of CO_2 emissions, but a short-lived hiccup in the economy did. In view of this, how can anyone deny that changing the economy must be part of the solution?

Looking at the bigger picture of the relationship of CO_2 emissions and the economy, we can take a look at CO_2 emissions during the past century (Figure 10.8). We can see a clear trend upward, but there are remarkable periods where emissions actually went down. All these periods are associated with events of great economic turmoil in the United States as well as worldwide, including both world wars and the Great Depression. We can also see right after World War II the steady climb of CO_2 emissions associated with the years of great economic growth in the United States. These are the years when "America became great" the first time around! This climb in atmospheric carbon is the price the planet had to pay for US economic well-being during those decades. These data bring the jury firmly in. We need to choose between the current economic system and the environment.

A couple of years ago I had a conversation with an older academic colleague, the great Harry Greene. Harry, like me, got his Ph.D. under the direction of Gordon Burghardt at the University of Tennessee. Although we did not overlap in graduate school, we are both alumni of Gordon's lab, so there always has been some sort of academic brotherhood between us. Harry had just given a presentation to the Colorado Partners in Amphibian and Reptile Conservation (CoPARC) where he talked about the possibility of

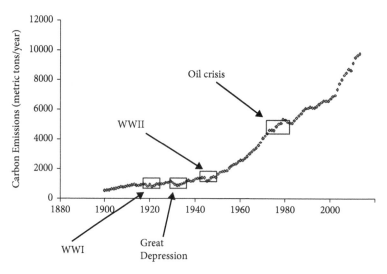

Figure 10.8 Emissions during the past century showing how major economic problems are associated with CO_2 emissions.[182]

rewilding some parts of North America with Asian elephants. He argued that it would restore ecological processes that existed in the continent before the extinction of a species of elephants very related to the Asian ones.[184] This would also allow us to protect Asian elephants from extinction stemming from the strong pressures in their natural habitats. In his presentation Harry referred to a conversation he had with someone skeptical about the notion of rewilding. Harry mentioned how toward the end of their conversation he presented the person with this choice (paraphrasing): "If it came down to it, what would you prefer, to keep the 'purity' of the North American prairies or to save the elephants?" The person eventually conceded that he would save the elephants.

Harry was signing copies of his book *Tracks and Shadows* while we were chatting about conservation issues following his lecture. Then the issue of capitalism as the source of the problems came up. To his resistance to consider changes in the dominant economic system, I presented Harry with the same conundrum he had presented his friend: "If you had to choose between changing capitalism or saving the elephants, what would you choose?"

His first response was to reject the question: "Why should we have to decide between those two?"

I pushed him further, using some of the earlier arguments I have presented about the incompatibility of consumerism and conservation. Some more people who wanted him to sign their copy of *Tracks and Shadows* interrupted our conversation, but after he signed another few copies there was a break. Harry looked at me sheepishly out of the corner of his eyes.

"You are still going to make me answer that question, aren't you?" he asked, not without a bit of resentment.

"Well, I will reword it to make it easier for you. What if by changing capitalism you will not only save the elephants, but you will also save the anacondas, and the rattlesnakes, and the polar bears, and the moths and caterpillars, and the birds, and all butterflies, and all the lizards, and dolphins, and whales, and the rest of earth's diversity. Wouldn't you change capitalism?" I asked him.

Of course, at this point the question did not need an answer. Harry replied the way most scientists do: "Change it for what? Change it how? I would not even know where to begin." This is certainly an insurmountable obstacle but, as I will discuss later, we need to rally together to give it our best try. We don't really have a choice!

The People Need Scientists to Be Honest

It is important that the scientific community is honest with the rest of society. Scientist know there are truly inviolable laws in nature. Thermodynamics is about the most unforgiving there is: *Energy cannot be created or destroyed; it can only be transformed.* So, scientists know that there is only so much you can produce with renewable energy. To be sure, there is a lot of room for us to improve the amount of renewable energy we can produce; there is much that technology can do. However, there is a hard limit that we cannot pass. We cannot propagandize our way to bend the laws of thermodynamics. It does not matter how much better we get in capturing renewable energy; there is always a limit to how much we can get.

On the other hand, industry can grow forever. There is no limit on how much industry can grow. The laws of economics require industries and factories to grow constantly. If renewables don't provide the energy needed, industry will resort to other sources. That is a fact. Those that grow more succeed; those that don't get driven out.

Again, let's put these two together. There is a hard limit on how much energy we can produce, but there is no limit on how much our industries can demand. At some point it is clear that no matter how good we get with renewables, there is a point at which they will not be enough. I believe it is the job of scientists to tell the lay audience the truth about this matter. We cannot power unlimited growth forever—even with fossil fuel and nuclear power. There will have to be a time where we must stop consumption. Whether it is now or a few decades down the road will determine how much biodiversity we have left and what is left of the planet we love.

Right now, there are millions of people out there hoping that science will solve conservation problems. There are countless people, educated, environmentally aware people, who are waiting for our guidance to tell them what to do. We need to tell them that we will not produce a solution and that no solution is possible in a system that is based on continued consumption. We need to come clean with the people or else we are failing our duty. For the scientific community to continue pretending that we can solve the problem with technology alone without telling the people the truth is akin of the captain of the *Titanic* telling everybody that they don't have to worry, even after he knew the ship was going down. It is irresponsible and borderline criminal.

The Lurking Monster

Public conservation figures have done a good job explaining to the world (at least to those who would listen) that global change is real and may produce unfathomable consequences. However, in their attempt to get people to like the conservation work, they neglected to tell the people something important: It will require sacrifices. Instead, they told the people to change their lightbulbs, buy more efficient washers, refrigerators, appliances, and cars—buy, buy, buy! It is all very good for business, but the increase in consumption can only hurt the environment. Of course, a campaign to lower consumerism will not be good for business. When people talk about the problem of global warming it always comes with the precondition that it "does not hurt the economy." We are told *ad nauseam* that the economy is the *bottom line*, the ultimate rule to live by. But scientists know that the true bottom line is the laws of thermodynamics. That is the real law that accepts no bargaining, no bending, and no negotiation. We need to live by it. Yet in the conservation discussions we act as if it is the economy represents the true unchangeable law and thermodynamics is just a "suggestion" we can push around at our will, when in actuality it is the other way around. The laws that rule the economy were made up by people, and we can change them and fix them as we please. They are often moved and shifted to help the powerful (as we did in the 2008 economic collapse), so they can also be changed to save the environment. It is all a matter of will.

No doubt I have exceeded the doses of iconoclastic rhetoric the reader is willing to bear, and I am also taking aim at the capitalist system. People are capable of great deeds. We are capable of deciphering the secrets of nature, and we have learned to predict the behavior of very complicated systems. However, there is something we do not do well: we do not unlearn well. Things that we have been taught, and on which we have based our jobs, lifestyles, society, and even strategies for conservation, are things that we have a hard time changing, or even considering that they may change. We just know that capitalism and the free market economy is the way to go. End of story. The greatest system. Period!

But the free market economy has failed in the very thing it is designed to do well: production of wealth. This might not be evident for the average US citizen because the United States is one of the few countries that has done relatively well within the free market system, but that is not the case for the majority of the countries that have tried it. When we look at the world in general, we have a handful of countries (basically the G8) that have done OK under the free market economy. On the other hand, we have the rest of the world (190-plus countries) that have done anywhere from not so well down to horribly bad. The likelihood of a country doing well within the free market economy can be calculated by dividing the few that have done well (8) by the number that have tried and failed (say, 180, excluding Cuba, China, and a handful of other countries that turned away from capitalism). The likelihood of a country succeeding in the free market economy is 0.04. Not significant!

Before September 2008, I felt the need to explain to the people that capitalism was not really sustainable and that relying on it for the preservation of diversity (or the running of the world, for that matter) was rather risky because the system was not reliable. The collapse of 2008 settled that debate for once and for all: capitalism is not sustainable. I am not saying that capitalism is not good for a sustainable world; capitalism is not sustainable itself! In 2008 we saw how it drove itself to self-destruction and had to be rescued using non-capitalist measures. Do we really want to entrust our precious diversity to a system so fickle that it stumbles and falls by itself? Notice that capitalism has failed most countries (and most people) in the accumulation of wealth, which is the one thing that capitalism is supposed to do well. When we look at the conservation of biodiversity, the wealthiest countries in the world are those that have destroyed most of their biodiversity and are the driving force for the destruction of biodiversity in other countries. We always hear that China produces more greenhouse gases than the United States. Yet never brought up in the discussion is the fact that the United States buys most of the products from China, so all those emissions we hold against China should actually be blamed on the developed world that buys the things that China produces. Lately the World Bank and IMF have hired some of the best ecologists in the world to produce literature and rather soft scientific work supporting the notion that the market economy is the best for the environment.[185] These studies amount to very serious mental gymnastics that manage to ignore the evidence to the contrary and conclude that a free market economy is the solution for conservation.[186] The adherence to the free market economy as a way to protect the environment is thus a truly ideological decision and is not based on data or reality.[155]

In the human psyche there is a lurking fear that our creativity can lead us to make something for our service that ends up being so good that it takes on a life of its own and ends up turning on us. That is the theme of the original story of Frankenstein. The great creation that Dr. Frankenstein brought back from the dead gets out of control and ends up hurting him. It is the same story-line of the *Terminator* movies, *2001: A Space Odyssey, Battlestar Galactica*, and countless other science fiction stories. The bad news is that that ship has already sailed. The creation of ours that has taken a life of its own and has turned against us is the economy. It was created to serve people, to settle matters of trade and bartering. It was supposed to be at the service of human-kind. But it has broken out of its cage, has taken a life of its own, and now rules the destinies of nations and peoples alike. The economy is no longer a tool to serve people. It has become a tyrannical master, and everyone must serve Her will. It bankrupts families and regions alike and leads people to commit horrible acts, even suicide. It starts wars and destroys entire coun-tries, not to mention devastating natural resources and the environment. It is time to pull out the pitchforks and torches and bring an end to our mon-strous creation.

So, I am convinced that we need to change our economic system if we want to make any progress in the conservation arena. Anything we try within the current economic system, which is directly responsible for the problem, is bound to fail. Clearly this is a reason for concern. Capitalism does offer great productivity and abundant commodities. Do we have to give all that up? Most people who have done between OK and well under capitalism will be very hesitant to the notion of changing it. Changing it how? Changing it for what? Don't change it too much! Nobody who has a winning hand wants other people to kick the table over.

Now this is the catch: I am not telling you that we need to quit capitalism and instead use "this" other system that works better. I am not proposing a new system, nor do I know what this new, better system looks like. I just know that capitalism will not, cannot, give us the solution because it is the source of the problem. I argue elsewhere that an economic system that is not based on the unlimited accumulation of capital, but on the well-being of the people, has a better chance of providing a framework for conservation efforts to flourish and obtain better results.[187] Whether we change and tweak capi-talism or replace it with something completely different is up to debate. But while we ignore this discussion and instead try to apply painkillers, we are only digging ourselves deeper and deeper into the hole.

Surely, someone will argue that the hiccup in the economy associated with lower CO_2 emissions produced lots of layoffs, more poverty, and more suf-fering among working-class families. However, I am not proposing that we have more and longer economic collapses to lower CO_2 emissions. Scientists can use the venom of very deadly snakes and produce miracle medicines to cure a variety of illnesses, including hypertension, cancer, and large list of medical ailments. They have not done this by injecting the raw venom into their patients. Rather, they have extracted the active ingredient that produces the desired physiological effect and administer it in the right doses. Can we extract from the economic collapse the active ingredient that lowers CO_2 emissions and prescribe it to fight climate change?

The data on climate change and the economy show unambiguously that the economy is the only thing that can make a difference, and it is foolhardy to leave it out of the equation. At this point all options should be on the table. The fact that we are not discussing alternative ways to change the economy that would allow us to solve the environmental crisis is because this discussion is dictated by strictly ideological dogmas and not by data, research, or interest in the planet. That will have to change before any progress takes place.

Lately some public figures in the United States have started to express the need to make real changes to address environmental problems. These proposals go beyond the incremental increase in fuel economy and energy generation. They are actually talking about real transformational changes in the way to run our society, including a different approach in the economic, social, and political realms (i.e., the Green New Deal). Although these are still incipient proposals and most deal with a general change of intent, rather hard and fast policies, I view these ideas as fresh air and I believe they may be the beginning of the change we really need. Of course, these public figures and their ideas have been ridiculed by status quo politicians and the mass media alike. Yet, this movement is a reason for optimism because, as we know, "there is something more powerful than the brute force of bayonets: it is the idea whose time has come and hour struck."[188, p. 57]

The main point I want to leave the reader with is that we must stop viewing politics and economics as if they were four-letter words wholly unrelated to our field. We must start learning about them and thinking how they relate to our conservation work. I want to create a point of equivalence,[194] or commonality, between people concerned with environmental degradation and loss of biodiversity and movements for sustainable economies (and perhaps anti-globalization movements) and get my fellow conservationists to realize that macroeconomics is simply another discipline in which we need to get involved with and which we need include in our environmental actions. Most conservationists do not hesitate to send an email to a state representative, sign a petition, or join a demonstration regarding issues such as conservation of roadless areas, joining international treaties for conservation, or searching for alternative ways to produce "green" energy. However, the same people are a lot more hesitant to support, in the same way, actions against policy brutality against minorities, structural racism, oppression of immigrants, Plan Colombia, economic embargoes and sanctions, regime change policies, policies of the World Trade Organization, the Free trade agreements, air strikes and other drone bombing programs, or other ways that the IEAs meddle with developing countries' sovereignties. Fellow conservationists, perhaps, do not see the relevance of it for conservation and do not fully understand the process. If I have succeeded in conveying the links among politics, economics, and conservation, this apathy toward macroeconomics on the part of biologists will recede. If we all look at the big picture when thinking of conservation problems, we can make more integral proposals that will have a better chance of success.

Ecology is a science that integrates many variables, seeking to understand their synergies and emergent properties of the systems. Isolating ecological problems from politics and economy goes against everything that matters in ecology. We cannot understand life and protect it if we close our eyes to the many variables affecting nature. Conservation of anacondas cannot be detached from conservation of everything else. I have tried to show that the current economic system is the cause of most of the conservation problems and any conservation efforts that do not include a change in the economic system are doomed to fail. However, no one said it better than the poet Wendell Berry: "We do need a 'new economy,' but one that is founded on thrift and care, on saving and conserving, not on excess and waste. An economy based on waste is inherently and hopelessly violent, and war is its inevitable by-product. We need a peaceable economy."[190]

11

Epilogue

A Different Approach

Caminante, son tus huellas
el camino y nada más;
Caminante, no hay camino,
se hace camino al andar.
Al andar se hace el camino,
y al volver la vista atrás
se ve la senda que nunca
se ha de volver a pisar.
Caminante, no hay camino
sino estelas en la mar.

(Traveler, it is your footprints
the only path, nothing else.
Traveler, there is no path;
you trace the path you walk.
As you walk, you trace the path,
and when you look back
you see the trail
that no traveler will walk on again.
Traveler, there is no path;
only wakes on the sea.)

Proverbios y Cantares XXIX, poem by Antonio Machado[191]

(Translated by JAR)

A Different Approach

I have shared my story of disentangling the secrets of the anaconda from my own point of view. I did not follow a prescription of what a book should contain, or proper ways of writing a book. A conventional format would have included a taxonomic review of the group in the first chapter, followed by anatomical and physiological information about the species, summarizing the existing knowledge from the literature, reproduction leading to the auto-ecology, population biology, and community interactions, ending in a standard approach to conservation based on sustainable use. A conventional style would be sterile and would not contain personal views, and all opinions would be carefully removed as nonscientific. Instead, I have shared with you my unapologetic ideas, mixed with scientific facts, opinions, and interpretations. I have done this in purpose because it's how actual humans think. I have not removed myself from my science because that would create the illusion that there is no bias in the science. Rather, I present my science with bias and all. In fact, I think that identifying my bias makes my science more reliable.

I believe this is important to contribute to the diversity of science. Trying to remove the human from the science renders the science stereotypical and hinders its ability to address a diversity of problems, because it does not have a diversity of views. Every picture has a camera angle, and to pretend that the picture does not have a point of view hinders our understanding of the scene. It is no different with science.

Here's a case in point. A conventional approach to discussing conservation would have emphasized the problems associated with human population growth and its resulting increased demand on natural resources and the other side of that coin: pollution. However, in the chapter about conservation I steered clear of any mention of population growth as a conservation problem. Instead, I focused on discussing the problems of consumerism and the links between the economy and the environment. In reality these two are very important issues in conservation. There are a couple of reasons that I chose to emphasize the problem of consumerism. One of them is because the reader can find people and scholars hammering *ad nauseam* on the problem of human population growth while the problem of consumerism is often ignored or not emphasized. So, I felt that it was more important that I focus on an area where more attention is needed.

The other reason I chose to emphasize consumerism as the main problem is because the problems of population growth and consumerism are not equally distributed geographically. Human population growth is a problem of developing/poor countries while consumerism is a problem of developed/rich countries. Most rich countries have naturally curbed their population growth as they become richer, so for the most part rich countries do not have a problem of population growth. In fact, many developed countries have started to experience decreases in their population because they have lowered the birth rate so much. On the other hand, poor countries, where population grows quickly, do not have a problem of consuming too many resources or producing too much waste. Putting it all together, if we had only a small population of humans consuming a lot, we would probably be OK. Also, if everybody had the same level of consumerism that poor countries have, we likely will not have an environmental crisis at all. So, both are very important problems, but the one that is often highlighted is the one that does not hurt businesses!

Also, there is a lot that rich countries can do to address the problem of consumerism. I, thus, focused my discussion on the influences of economics in conservation where I believe rich countries can make a more important contribution to the problem. On the other hand, there is very little that rich countries can do to address the problem of population growth. In fact, the only thing that rich countries can do about it is tell brown people in poor countries that they are the problem. Clearly this is not a very fruitful strategy for conservation.

Thus, this book, untraditional and unexpected as it may be, reflects my view of the world. I didn't follow a proper, prescribed path to become a scientist; in fact, the way I have developed my scientific career has been anything but traditional or expected. It comes as no surprise, then, that the way I carry out my research is also unique and unconventional.

Study Site: A Moving Target

Most field studies are done in a "study site," normally a field station or a piece of land that can be identified with one name and/or one owner. This was far from the case in my study. Earlier on, I mentioned that most of my research was done in Hato El Cedral, a cattle ranch located in Apure State. However, this is just the name it had when I started. In those days it was owned by Agropecuaria Rio Yaracuy, a cattle company that had properties all over the country as well as various investments in other fields, including banks, insurance companies, and much more. At this time the company also had a substantial tourist operation that generated a nice revenue on top of the cattle, which was the main business.

Starting in the 1980s and into the 1990s Venezuela was interested in exploiting wildlife sustainably. This is part of the reason that Profauna got onboard with funding my project. At this time sustainable use of caimans was well under way, and a lot of the Profauna operations were being funded by the tax revenue the program created. El Cedral had built facilities to raise spectacled caimans for commercial use. It was an open farm where they collected the eggs from the wild, incubated them in captivity, and raised the babies for 1 year. After this time the babies were harvested for their skin, and a portion of them were released into the wild. The pen was a circular cement enclosure, about 5 meters (16.4 feet) in diameter, with a pit in the middle containing water, surrounded by a tin sheet (Figure 11.1). The tin sheet was about 50 centimeters (20 inches) high, and above that there was a chain-link fence up to 1.5 meters (5 feet) or so. The roof of the enclosure was also chain link to prevent hawks and aerial predators from taking the baby caimans. My project benefited greatly from these facilities, since I used the pens to hold pregnant anacondas after the caiman farm went out of business (see Chapter 10). But while the farm was working the manager, Saul Gutierrez, who was a fan of crocodilians, acquired a young Orinoco crocodile. The croc had been a pet in someone's house and after it surpassed 1.2 meters (4 feet) the owners could no longer have her, knowing that she would eventually become almost that wide!

Saul accepted the croc and there was a big discussion about what to name her. "Name her Mónica," "Name her Cecilia," "Name her Teresa": Just about every man who worked in the station wanted the croc named after his wife. Eventually the croc was named "Mi Mujer" (My Wife)! Mi Mujer lived there with a pen for herself since she was much larger than the baby caimans that were raised in the facility. As the facilities were a few years old, John, as well as María and myself, noticed that the pen that held Mi Mujer had some holes in the upper part of the fence. We pointed it out to Saul but because it was high, he did not think it would be an escape hazard and Mi Mujer was much too big to worry about aerial predators. We pointed out to him on repeated occasions that crocs can climb a fence like that, but our warning fell on deaf ears.

At the end of my second season working with anacondas, I started my Ph.D. program at the University of Tennessee and left my field site, where I had lived for 1.5 years. I packed my luggage and left with the intention of coming back the next year in the dry season to continue my study. Unfortunately, my departure occurred between the last time Mi Mujer was seen and the time her cage was found empty. The logical connection was that I had stolen the caiman before I departed. Who else would steal a caiman that size? To make matters worse, a friend of mine, Pedro Azuaje, who had a small conservation farm 3 hours away, had acquired a young caiman about the same size as Mi Mujer right about the same time that Mi Mujer disappeared. So, obviously it was me who had stolen a 1.5-meter-long (5-foot-long) caiman in the dead of night, had smuggled it among my gear in the truck, and had given it my friend in the other farm. Because Mi Mujer had no tags or marking of any sort, there was no way to prove my innocence one way or the other.

Ignorant of all this, I contacted the ranch the next year, notifying them of my intentions to come back to resume my study. The answer came back that I was not welcome at the ranch because I had stolen Mi Mujer. John Thorbjarnarson, who was a member of my committee and had been involved with me from the beginning of the study, was told that he could come back and my assistant María could also come back if she wanted (although presumably she would have been my accomplice in stealing Mi Mujer). This would have been the end of the study, though. John tried to convince the owners that I had not stolen the croc, but they were dead set against letting a thief back on their property. In the meantime John heard from Victor Delgado, a fishermen who worked in the ranch, that he had seen a small crocodile in a creek (Caño Guaratarito) not 2 kilometers (1.2 miles) from the place where the Mi Mujer's pen was, the closest body of water to her pen!

John set out to try to catch the crocodile to prove my innocence. Since his specialty was precisely crocodiles, he succeeded after a couple of days trying and brought the tied-up 1.7-meter-long (5.6-foot-long) crocodile to the ranch managers.

"That is not Mi Mujer," responded the station manager. "That croc is larger than she was."

Then John went to the literature and showed them that 20 centimeters (7.9 inches) was about the growth expected in a crocodile her age during the time that Mi Mujer had been missing. Eventually, the owners grudgingly agreed to let me back on the ranch toward the very end of the dry season, so I could at least get some data for that year.

That I could come back and continue the study is one part of the happy ending of the story. The other part is that toward the end of that year some of the investments of Agropecuaria Rio Yaracuy collapsed. They borrowed many millions from the government to invest in rescuing their company, but instead they embezzled the money and moved to the United States to start another venture capital business there. Eventually they embezzled some money in the United States too, and they got caught, landing themselves in jail. The person who accused me of stealing the crocodile, his father, and his son all ended up doing time in a federal prison. Then everybody knew who the real thieves were!

As the owners were in jail, the ranch was auctioned and ended up in the hands of another cattle ranching conglomerate called COVEGAN that had even more properties all over the country. Eventually the Venezuelan government bought a controlling share of the ranch and made it into a production ranch, applying concepts of sustainable use combined with traditional ranching.

Passing the Baton

Studying the life of a species as long-lived and broadly distributed as anacondas is a major undertaking. I am glad I've gotten as far as I have, but I am nowhere close to the end. There is still lot more to study about many aspects of their biology and in many regions. So far, I have shed light on the life of anacondas in Los Llanos, but given the vastness and complexity of the different habitats they inhabit, there is clearly much to be learned about them. So far, I have learned that a lot of what I know applies to other locations but also that there are enough differences to make other studies very worth undertaking. Being in my mid-50s, I like to think that I have another good 10 years in which I can still tangle up with a big snake, and while it is true that nowadays I use a lot more brains than brawn when dealing with a big snake, it seems that as I approach 65, I should pass the baton to the younger generations. I have been able to lend my help and expertise to other scientists to start collaborative projects in other parts of South America that hopefully will yield also valuable information.

The Value of Diversity

As I look at my contribution to my discipline, I feel that what I have done has been a direct consequence of my unique and personal approach to doing science. If I had tried more conventional approaches, I might have learned interesting stuff, but not what I have learned. Putting this into a bigger context, this is an important component of diversity in the sciences.

In the face of the current extinction crises, we hear a lot about protecting biological diversity and the value of diversity for the world and for future generations and to protect evolutionary processes. Also, as we live in a more and more polarized world where resources are becoming scarce, we see the resurgence of intolerant attitudes toward minorities—ethnic, religious, gender—that we thought had died out decades ago. Often those concerned with biological diversity do not understand, or care, about the other meaning of diversity and those interested in cultural, ethnic, or sociological diversity know little about biological diversity. Should these two parties have a conversation, one might hear a remark like, "Oh, no, that isn't what I mean by 'diversity'!" I want to take a minute to reconcile those two meanings and show that they are not that different in the bigger picture.

As a foreign-born, Hispanic person who teaches in a public university, I am a member of a very small minority of faculty in the United States. Some people see this as a feather that my university administrators can put in their cap for hiring minorities and promoting diversity. But what does my presence as a faculty member mean for the university? Most university professors were good students all their lives, graduated in 4 years from college with a high grade-point average (GPA), and then obtained graduate degrees by following instructions and procedures, all of which were neatly laid out for them. They jumped through hoops during their graduate programs, applied for grants in the right places, and did the postdocs that worked best for their career. At the end of this path was a tenure-track position at a university.

My path was slightly different. I attended Universidad Central de Venezuela, where I obtained the best education money can buy, but it cost me nothing. Because of this it took me 9 years to obtain my undergraduate degree. I was busy exploring the world while I was in college. I was a firefighter; I ran; I swam; I did scuba diving, rock climbing, spelunking, all kinds of hiking, and just about anything I found fun. I climbed the ranks in the fire department and became the commander during my last year of service. During a lot of this time I took about two courses per semester and was not very focused on my classes unless I loved the subject. In Venezuela grades are in a 0-to-20 scale. My graduate average was 14.2, what would be a C in the United States. It was just 0.2 above the average of my class, composed of several hundred graduates.

What did it mean when I applied to a very nondiverse US school system? Clearly, spending 9 years in an undergraduate career raised eyebrows and a GPA of C did not win me any friends in the admissions committee. Luckily, I did not apply for graduate school right after I earned my undergraduate degree. Instead I did quite a bit of research. Stuart Strahl was the assistant director for Latin American programs of the New York Zoological Society (currently the Wildlife Conservation Society) at the time and resided in Venezuela. Stuart knew me and my love for animals and field research. I shared with him my interest in continuing to study iguanas that had originated with my undergraduate thesis.[192] He was able to fund my expenses in the field. This was also a very unorthodox approach to research. It was not a job; I didn't work as somebody's assistant the way most recent graduates do and I did not get a salary. Being young and uncommitted, all that I needed were basic field research expenses, food, and a place to hang my hammock. To add to how unconventional it all was, there were rumors that Stuart did not fund these kinds of projects with funding from the Wildlife Conservation Society but with his own money, just because he thought it was a good cause and wanted to help out. I was able to do a few years of research with iguana demography and more importantly iguana behavior. I discovered that male baby iguanas attract predators toward themselves to help their sisters survive.[193] When I applied to the Ph.D. program at the University of Tennessee, I wrote Dr. Gordon Burghardt about my finding and my interest in doing this research under his supervision, since he had studied baby iguanas himself. Gordon's positive response was critical: "Your iguana work sounds fascinating and I would certainly consider you as a graduate student." Gordon lobbied to get me accepted into the program despite much opposition from the conventional faculty. By this time, I had already begun my study on anacondas working (now a real job) for Profauna, but Gordon was unaware of it when he accepted me!

The objection of the admissions committee was well justified. I did not conform to the kind of students who succeed in graduate school in the United States. But as it happens, part of my success was due to my unorthodox approach to science. When I was doing the work with baby iguanas, there was a consensus in the herpetological community that it was not possible to tell the sex of babies I had not read the literature that presented this view, partially because of my inability to find up-to-date literature in a Venezuelan university at the time. When I was studying iguanas to learn about their diet, I kept inspecting their cloacas to see if I could tell the sex among babies. While doing this I learned that there is a clear difference in the shape of their cloaca that allows one to unambiguously tell a boy from a girl.[194] This was something that people did not think was possible and had not tried! This breakthrough allowed me to do the research that eventually caught Gordon's attention. Had I read the literature saying that it was not possible to tell the sex among neonatal iguanas, I would have never tried and my research would not have started. I am not saying that scientists should be ignorant on their subject matter, but I am saying we need to be open-minded. Trying things that everybody knows "won't work" is not always a bad idea.

What does it matter that I got into my Ph.D. program by the skin of my teeth? Eventually I got a Ph.D. like everybody else. Well, while I was in my Ph.D. program, I had plenty of opportunities to follow a standard path of research like everybody else. But I continued to try something that I found better suited to my own interests. When I turned in my research proposal to my doctoral committee, the title "Natural History of the Green Anaconda" resulted in a strong objection from some of the members: "Natural history is not science!" The fad in academia during the past few decades has been to frown on natural history; it is considered an obsolete way of doing science and no longer scientific enough for today's standards. Fortunately, other members of the committee (who had read my proposal more carefully) could tell that there was a lot of good science being proposed; I was just not doing the dance and using the buzzwords that were needed for my proposal to fit in. Thanks to them, though, my proposal was finally accepted.

Up to the time I started to study anacondas, many a herpetologist had thought of doing it but likely did not think it was possible. My approach to doing my study was not in the books, and anyone trying to learn how to study anacondas, at the time, would have not found any guidance. A conventional scientist changes tack for something more "doable" that will produce results in the expected time, with the allotted funding. This is not what I was doing. I was not interested in doing a conventional successful study, and I dared to try something else. Because of all my "wasted" time during my undergraduate education, I had developed an understanding of the local folks in the llanos. I knew their lingo and could speak with them in their terms. This was critical when I tried to learn from Don Pedro, Ramón Arbujas, Victor Delgado (el Viejo), and others. I did not have to be the smart one in the group. I respected their older age and sought their knowledge. They understood that and shared with me their wisdom. This is the value of diversity. Thanks to my unconventional approach to science, I tried something different and I learned something new.

While some people may view the inclusion of ethnic and other kinds of diversity in science as "tokenistic" and unnecessary, diversity is as important in science as it is in the preservation of biodiversity. It is important for populations to have genetic diversity because it provides a defense against environmental unpredictability. It is also important because it provides the foundations for future evolutionary processes. This is just as true with scientific approaches. Science is only as good as the question, and we cannot learn the answers to questions we don't ask. We need to increase the range of questions we ask, and that comes from having people with diverse backgrounds to ask those questions. We have seen great improvement in the quality of science since women started participating more actively in scientific endeavors. Their contributions have been important in the whole gamut of fields, from the discovery of the structure of DNA to new approaches to study the natural history of wildlife. We need more of that, and we need more of more. Not only do we need to ask questions from our highly educated, often privileged perspective, but we also need to ask questions from the point of view disenfranchised, the poor, and the down-trodden; the oppressed, and the unfitting. We need questions from those of brown and mixed heritage. We need questions from the non-binary, the gay, and the "queird." Only with a diversity of questions we get a diversity of learning. Only with diversity of science can we evolve in new directions, attending to the needs of a changing world. Diversity of backgrounds and diversity of experiences allow for that diversity of questions and answers that will allow our scientific thinking to evolve and move forward into new horizons.

A New World

I was surprised that I came to the United States to study. I never thought I would leave Venezuela. Once I graduated, I had lots of animals to study and lots of time to do it. I would have never sought to leave the country if it hadn't been that my mentor in the study of wildlife, Juhani Ojasti, was retiring. I had looked up to him since I started my career, and he always gave me good advice even if it was not his job to help me. However, after 30-plus years teaching. he felt the need to return to Finland, his country of origin. After Ojasti retired there was no one in the country I saw as my mentor. Stuart had returned to the United States working for Audubon, John Thorbjarnarson moved to work in Brazil, and Luis Levín, the only one left in Venezuela, re-tired from teaching.

After my miraculous acceptance to the University of Tennessee I discov-ered a new world. I had always been the local, so being the foreigner was quite a change. I barely spoke the language, and being a Hispanic person in Knoxville, I stood out like a sore thumb and was way out of my element. I thought I knew enough English to get by, but all that went away the first time I met a local in an off-campus store in Knoxville! It was quite a learning experience, to say the least.

Over the years I have heard many times people praising my courage for working with anacondas and large dangerous reptiles. I never saw that as courage; it's more like excitement, interacting with a wonderful animal I ad-mire. However, leaving my homeland and moving to Tennessee required every inch of courage I could muster: I was in a different country, with a dif-ferent culture, speaking a foreign language, away from my friends, family, and loved ones, in a place where I did not belong. For those who brave leaving their homelands to explore a new world, I give them a firm tip of my hat for their incredible courage.

After studying a couple of bear cubs for a few years, my Ph.D. advisor, Gordon Burghardt, wrote as he contemplated the distance between the world of the bears and his own: "The affective world of the black bears does not mirror ours. Nor is it a parallel alien universe, but one running oblique to or own. If we have the privilege of being individually bonded to bears, their world touches ours at their convenience, and we are given partial access to it through a darkened glass, but never face to face."[195, p. 379] Gazing at the abyss between my world and that of the anacondas I feel even further removed. Our phylogenetic similarities to bears give us far more context than we can ever have with snakes. Do they even have an affective world? What is it like? How is it expressed? Can we relate to it? After all these years I still feel very far from understanding their world as well as I would like and am still un-able to look into their world as if it were mine. I am limited to just a snapshot through murky water between us. Yet I feel equally privileged to have had the opportunity to explore their world, navigate their ways, and peek into their secret lives so far removed from ours.

If I have succeeded with this book, I hope I have taught the reader a thing or two about anacondas, but also I hope I conveyed my story and how I went about learning what I have learned. I hope the reader leaves these pages feeling that there are many ways to do science and many ways to make a difference. Struggle and strife are part of life in any field one chooses, and one does not get anywhere without facing at least some hurdles and problems. At the beginning I did not have any special head start to succeed; all I had was the right motivation. It is true that I was very lucky to find the right people at the right time to help me, but luck alone would not have gotten me very far. While there is no fail-proof prescription for success, there is one for failure: not trying.

Literature Cited

1 Gumilla, J., *El Orinoco ilustrado y defendido* (Editorial Arte, Madrid/Caracas, 1740/ 1963).

2 Rivas, J. A., Conservation of anacondas: how Tylenol conservation and macroeconomics threaten the survival of the world's largest snake. *Iguana* 14 (2), 10 (2007).

3 Rivas, J. A. and Burghardt, G. M., Crotalomorphism: a metaphor to understand anthropomorphism by omission, in *The Cognitive Animal: Empirical and Theoretical Perspective on Animal Cognition*, edited by M. Bekoff, C. Allen, and G. M. Burghardt (MIT Press, Cambridge, MA, 2002), p. 9.

4 Rivas, J. A., et al., Natural history of the green anacondas in the Venezuelan llanos, in *Biology of Boas, Pythons, and Related Taxa*, edited by R. W. Henderson and R. Powell (Eagle Mountain Publishing, Eagle Mountain, UT, 2007), p. 128.

5 Rivas, J. A., Rodriguez, J. V., and Mittermeier, C. G., The llanos, in *Wilderness*, edited by R. Mittermeier (Cemex, Mexico, 2002), p. 265.

6 Greene, H. W. and Hardy, D. L., Natural death associated with skeletal injury in the f. *Copeia* 1989 (4), 1036 (1989).

7 Rivas, J. A., et al., A safe method for handling large snakes in the field. *Herpetological Review* 26, 138 (1995).

8 Rivas, J. A., What is the length of a snake? *Contemporary Herpetology* 2008 (2), 1 (2008).

9 Gallegos, R., *Doña Barbara* (Domingo Milani, Madrid, Spain, 1929).

10 Rivas, J. A., How I survived killer bees, in *Complete Survival Manual*, edited by M. Sweeney (National Geographic, Washington, DC, 2009), p. 68.

11 Rivas, J. A., Molina, C. R., and Ávila, T. M., A non-flushing stomach wash techniques for large lizards. *Herpetological Review* 27 (2), 72 (1996).

12 López-Corcuera, G., *Fauna legendaria: rayas, tembladores, caimanes y culebra de aguas* (Editorial Arte, Fundación Cientifica Fluvial de los Llanos, Caracas, Venezuela, 1984).

13 Rivas, J. A., Molina, C. R., Corey-Rivas, S. J., and Burghardt, G. M., Natural history of neonatal green anacondas (*Eunectes murinus*): A chip off the old block. *Copeia* 104 (2), 402 (2016).

14 Rivas, J. A., Owens, R. Y., and Calle, P. P., *Eunectes murinus*: juvenile predation. *Herpetological Review* 32 (2), 107 (2001).

15 Rivas, J. A., *Life History of the Green Anaconda with Emphasis on Its Reproduction Biology* (CreateSpace Independent Publishing Platform, North Charleston, SC, 2015).

16 Calle, P., et al., Health assessment of free-ranging anacondas (*Eunectes murinus*) in Venezuela. *Journal of Zoo and Wildlife Medicine* 25, 53 (1994).

17 Morafka, D. J., Spangenberg, E. K., and Lance, V. A., Neonatology of reptiles. *Herpetological Monographs* 14, 353 (2000).

18 Rivas, J. A., Muñoz, M. de C., Burghardt, G. M., and Thorbjarnarson, J. B., Mating system and sexual size dimorphism of green anaconda (*Eunectes murinus*), in *Biology of Boas, Pythons, and Related Taxa*, edited by R. W. Henderson and R. Powell (Eagle Mountain Publishing, Eagle Mountain, UT, 2007), p. 461.

19 Rivas, J. A. and Burghardt, G. M., Sexual size dimorphism in snakes: wearing the snake's shoes. *Animal Behaviour* **62** (3), F1 (2001).

20 Savitzky, B. A. and Burghardt, G. M., Ontogeny of predatory behavior in the aquatic specialist snake, *Nerodia rhombifer*, during the first year of life. *Herpetological Monograph* **14**, 401 (2000).

21 Greene, H. W., The ecological and behavioral context for pitviper evolution, in *Biology of Pitvipers*, edited by J. A. Campbell and E. D. Brodie (La Selva, Tyler, TX, 1992), p. 107.

22 Sazima, I., Natural history of the jararaca pitviper, *Bothrops jararaca*, in southern Brazil, in *Biology of Pitvipers*, edited by J. A. Campbell and E. D. Brodie (La Selva, Tyler, TX, 1992), p. 199.

23 Branch, W. R. and Haacke, W. D., A fatal attack on a young boy by an African rock python (*Python sebae*). *Journal of Herpetology* **14**, 305 (1980).

24 Rivas, J. A., Predatory attack of a green anaconda (*Eunectes murinus*) on an adult human. *Herpetological Natural History* **6**, 158 (1998).

25 Shine, R. G., Harlow, P. S., Keogh, J. S., and Boeadi, The influence of sex and body size on food habits of a giant tropical snake, *Python reticulatus*. *Functional Ecology* **12**, 248 (1998).

26 Andreadis, P. T. and Burghardt, G. M., Unlearned appetite controls: watersnakes (*Nerodia*) take smaller meals when they have the choice. *Journal of Comparative Psychology* **119**, 304 (2005).

27 Lange, A., *The Lower Amazon* (G. P. Putnam's Sons, New York, 1914).

28 Rivas, J. A., *Eunectes murinus* (green anaconda): subduing behavior. *Herpetological Review* **35** (1), 66 (2004).

29 Boback, S. M., et al., Snake modulates constriction in response to prey's heartbeat. *Biology Letters* **8** (3), 473 (2012).

30 Valderrama, X. and Thorbjarnarson, J. B., *Eunectes murinus*: predation. *Review* **32**, 46 (2001).

31 Secor, S. M. and Diamond, J., Determinants of the postfeeding metabolic response of Burmese pythons, *Python molurus*. *Physiological Zoology* **70**, 202 (1997).

32 Strussman, C., Hábitos alimantares da sucuri-amarela, *Eunectes notaeus* Cope, 1862, no pantanal Matogrossense. *Biociências* **1**, 35 (1997).

33 De Freitas, D., *Eunectes murinus* (green anaconda): diet. *Herpetological Review* **40** (1), 98 (2009).

34 Rivas, J. A. and Owens, R. Y., *Eunectes murinus* (green anaconda) cannibalism. *Herpetological Review* **31** (1), 44 (2000).

35 O'Shea M.T., *Eunectes murinus gigas* (Northern green anaconda) cannibalism. *Herpetological Review* **25**, 124 (1994).

36 Shine, R. G., Ambaiyanto, Harllow, P. S., and Mumpuni, Ecological attributes of two commercially-harvested python species in northern Sumatra. *Journal of Herpetology* **33**, 249 (1999).

37 Headland, T. N. and Greene, H. W., Hunter–gatherers and other primates as prey, predators, and competitors of snakes. *Proceedings of the National Academy of Sciences* **108** (52), E1470 (2011).

38 Stephen, D. W. and Krebs, J. R., *Foraging Theory* (Princeton University Press, Princeton, NJ, 1986).

39 Oviedo-y-Baños, J., *The Conquest and Settlement of Venezuela* (University of California Press, Berkeley, CA, 1987), p. 1723.

40 Bell, G., The costs of reproduction and their consequences. *The American Naturalist* **116**, 45 (1980).

41 Daan, S. and Tinbergen, J. M., Adaptation of life history strategies, in *Behavioural Ecology: An Evolutionary Approach*, edited by J. R. Krebs and N. B. Davies (Blackwell Science, Oxford, UK, 1997), p. 311.

42 Williams, G. C., Natural selection, the costs of reproduction, and a refinement of Lack's principle. *American Naturalist* **100**, 687 (1966).

43 Kozlowski, J. and Wiegert, R. G., Optimal allocation of energy to growth and reproduction. *Theoretical Population Biology* **29**, 16 (1986).

44 Hamilton, W. D., The genetical evolution of social behaviour. *Journal of Theoretical Biology* **7**, 1 (1964).

45 Arnold, S., Foraging theory and prey-predator-size in snakes, in *Snakes: Ecology and Behavior*, edited by R. A. Seigel and J. T. Collin (McGraw-Hill, New York, 1993), p. 87.

46 Shine, R., Relationship among body size, clutch size, and egg size in three species of oviparous snakes. *American Naturalist* **134**, 311 (1989).

47 Ford, N. and Seigel, R., Relationship among body size, clutch size, and egg size in three species of oviparous snakes. *Herpetologica* 1989, **45**:75-83.

48 Sinervo, B. and Licht, P., Proximate constraints on the evolution of egg size number and total clutch mass in lizards. *Science* **252**, 1300 (1991); Stearns, S. C., *The Evolution of Life Histories* (Oxford University Press, New York, 1992).

49 Pianka E. R., On r- and K-selection. *American Naturalist* 1970, **104**.

50 Wiewandt, T. A., Evolution of nesting patterns in iguanine lizards, in *Iguanas of the World: Their Ecology, Behavior and Conservation*, edited by G. M. Burghardt and A. S. Rand (Noyes Publications, Saddle River, NJ, 1982), p. 119.

51 Lack, D., Ecological adaptation for breeding in birds, in *Ecological Adaptation for Breeding in Birds* (Methuen and Co. Ltd., London., 1968) .

52 Tinbergen, N., On aims and methods of ethology. *Zeitschrift für Tierpsychologie* **20**, 410 (1963).

53 Burghardt, G. M., Amending Tinbergen: a fifth aim for ethology, in *Anthropomorphism, Anecdotes, and Animals*, edited by R. W. Mitchell, N. S. Thompson, and H. L. Miles (SUNY Press, Albany, NY, 1997), p. 254.

54 Fulton, T. W., The rate of growth of fishes, in *22nd Annual Report* (Fishery Board of Scotland, 1904), Vol. 3, 141; Ricker, W. E., Computation and interpretation of biological statistics of fish populations. *Bulletin of the Fisheries Research Board of Canada* **191**, 1 (1975).

55 Schuett, G. W., Is long term sperm storage an unimportant component of the reproductive biology of temperate pitvipers? in *Biology of Pitvipers*, edited by J. A. Campbell and E. D. Brodie (La Selva, Tyler, TX, 1992), p. 169.

56 Holmstrom, W. F. and Behler, J. L., Post-parturient behavior of the common anaconda *Eunectes murinus. Zoological Garten N. F.* **51**, 353 (1981).

57 Belluomini, H. E. and Hoge, A. R., Operaçao cesariana realizada em *Eunectes murinus* (Linnaeues 1758) (Serpentes). *Memorias do Instituto Butantan* **28**, 187 (1957/58).

58 Martin, W. H., Phenology of the timber rattlesnake (*Crotalus horridus*) in an unglaciated section of the Appalachian Mountains, in *Biology of Pitvipers*, edited by J. A. Campbell and E. D. Brodie (La Selva, Tyler, TX, 1992), p. 259.

59 Harlow, P. and Grigg, G., Shivering thermogenesis in a brooding diamond python, *Python spilotes spilotes. Copeia* **1984**, 959 (1984).

60 Hutchinson, V. H., Dowling, H. G., and Vinegar, A., Thermoregulation in a brooding female Indian python (*Python molurus bivittatus*). *Science* **151**, 694 (1966).

61 Ross, R. A. and Marzec, G. M., *The Reproductive Husbandry of Pythons and Boas* (Institute of Herpetological Research, Stanford, CA, 1990).

62 Ford, N. B., The occurrence of anomalies in broods of *Thamnophis* in relation to female body size. *Occasional Papers of the Dallas Herpetological Society* **2**, 1 (1980).

63 Rivas, J. A., Thorbjarnarson, J. B., Owens, R. Y., and Muñoz, M. d. C., *Eunectes murinus*: caiman predation. *Herpetological Review* **30** (2), 101 (1999).

64 Rivas, J. A. and Corey, S. J., *Eunectes murinus* (green anaconda). Longevity. *Herpetological Review* **39**, 469 (2008).

65 Burghardt, G. M., Chemical prey preference polymorphism in newborn garter snakes, *Thamnophis sirtalis*. *Behaviour* **52**, 202 (1975).

66 Pizzatto, L., Madsen, T., Brown, G. G., and Shine, R., Spatial ecology of hatchling water pythons (*Liasis fuscus*) in tropical Australia. *Journal of Tropical Ecology* **25**, 181 (2009).

67 Seigel, R. A. and Ford, N., Reproductive ecology, in *Snakes: Ecology and Evolutionary Biology*, edited by R. A. Seigel, J. T. Collin, and S. S. Novak (McGraw-Hill Inc., New York, 1987), p. 221.

68 Madsen, T. and Shine, R. G., Determinants of reproductive output in female water pythons (*Liasis fuscus*: Pythonidae). *Herpetologica* **52**, 146 (1996).

69 Shine, R. G., Sexual size in snakes revisited.*Copeia* **1994**, 326 (1994).

70 Shine, R. G., Sexual dimorphism in snakes, in *Snakes: Ecology and Behavior*, edited by R. Seigel and J. T. Collin (McGraw-Hill Inc., New York, 1993), p. 49.

71 Seigel, R. A. and Fitch, H. S., Ecological patterns of relative clutch mass in snakes. *Oecologia* **61**, 293 (1984).

72 Slip, D. J. and Shine, R. G., The reproductive biology and mating system of the diamond python, *Morelia spilota* (Serpentea: Boidae). *Herpetologica* **44**, 396 (1988).

73 Stewart, J. R., Backburn, D. G., Baxter, D. C., and Hoffman, L. H., Nutritional provision to embryos in a predominantly lecithotropic placental reptile *Thamnophis ordinoides*. *Physiological Zoology* **63**, 722 (1990).

74 Huxley, J., *Problems of Relative Growth* (Methuen, London, 1932).

75 Zeh, J. A. and Zeh, D. W., The evolution of polyandry I: intragenomic conflict and genetic incompatibility. *Proceedings of the Royal Society of London Series B* **263**, 1711 (1997).

76 Whittier, J. M. and Crews, D., Mating increases plasma levels of prostaglandin F2α in female garter snakes. *Prostaglandins* **37**, 359 (1989).

77 Ford, N. and O'Blesness, M. L., Species and sexual specificity of pheromone trails of the garter snake, *Thamnophis marcianus*. *Journal of Herpetology* **20**, 259 (1986).

78 Ford, N. and Low, J. R., Sex pheromone source location by garter snake: a mechanism for detection of direction in nonvolatile trails. *Journal of Chemical Ecology* **10**, 1193 (1984).

79 Madsen, T. and Shine, R. G., Determinants of reproductive success in female adders, *Vipera berus*. *Oecologia* **92**, 40 (1992).

80 Shine, R., Harlow, P. S., Keogh, J. S., and Boeadi, The allometry of life-history traits: insigths from a study of giant snakes (*Python reticulatus*). *Journal of Zoology* **244**, 405 (1998).

81 Ford, N. B. and Seigel, R. A., An experimental study of the trade-offs between age and size at maturity: effects of energy availability. *Functional Ecology* **8**, 91 (1994).

82 Ford, N. B. and Siegel, R. A., Effect of energy input on variation in clutch size and off-spring size in a vivparous reptile. *Functional Ecology* 6, 382 (1992).

83 Riechert, S. E., A test for phylogenetic constraints on behavioral adaptation in a spider system. *Behavioral Ecology and Sociobiology* 32, 343 (1993).

84 Riechert, S. E., Investigation of potential gene flow limitation of behavioral adaptation in an aridlands spider. *Behavioral Ecology and Sociobiology* 32, 355 (1993).

85 Topsel E: *The History of Four Footed Beast, Serpents, and Insects* (E. Cotes, London, 1658).

86 Beebe, W., Field notes on the snakes of Kartarbo, British Guiana and Caripito, Venezuela. *Zoologica* 31, 11 (1946); Blomberg, R., Giant snake hunt. *Natural History* 1956, 92 (1956); Pope, C. H., *The Giant Snakes* (Alfred Knopf, New York, 1961); Gilmore, R. M. and Murphy, J. C., On large anacondas, *Eunectes murinus* (Serpentes: Boidae), with special reference to Dunn-Lamon record. *Bulletin of the Chicago Herpetological Society* 28, 185 (1993).

87 Pritchard, P. C., Letters to the editors. *Bulletin of the Chicago Herpetological Society* 29, 37 (1994).

88 Head, J. J., Bloch, J. I., Hastings, A. K., Bourque, J. R., Cadena, E. A., Herrera, F. A., Polly, P. D., and Jaramillo, C. A., Giant boid snake from the Palaeocene neotropics reveals hotter past equatorial temperatures. *Nature* 2009, 457:715–717.

89 Arnold, S. and Duvall, D., Animal mating system: a synthesis based on selection theory. *The American Naturalist* 143, 137 (1994).

90 Rivas, J. A. and Burghardt, G. M., Snake mating system, behavior and evolution: the revisionary implications of recent findings. *Journal of Comparative Psychology* 119 (4), 447 (2005).

91 De La Quintana, P., Pachecho, L., Rivas, J. A., *Eunectes beniensis: Cannibalism*. *Herpetological Review* 2011, 42(4): 614.

92 Mason, R. T. et al., Sex pheromones in snakes. *Science* 245, 290 (1989); Mason, R. T., Reptilian pheromones, in *Biology of the Reptilia*, edited by C. Gans and D. Crews (University of Chicago Press, Chicago, 1992), p. 114; Ford, N. and Holland, D., The role of pheromone trails in spacing behaviour in snakes of snakes, in *Chemical Signals in Vertebrates 5*, edited by D. MacDonald, D. Muller-Schwarze, and S. E. Natynczuk (Oxford University Press, Oxford, 1990), p. 465.

93 Andersson, M., *Sexual selection*. (Princeton University Press, Princeton, NJ, 1994), p. 599.

94 Schuett, G. W. et al., Production of offspring in the absence of males: Evidence for facultative parthenogenesis in bisexual snakes. *Herpetological Natural History* 5 (1), 1 (1997).

95 Booth, W. et al., Evidence for viable, non-clonal but fatherless *Boa constrictors*. *Biological Letters*. doi:10.1098/rsbl.2010.0793 (2010).

96 Parker, G. A., Sperm competition and its evolutionary consequences in the insects. *Biological Review* 45, 525 (1970).

97 Gillingham, J. C., Carpenter, C. C., and Murphy, J. B., Courtship, male combat and dominance in the western diamond back rattlesnake *Crotalus atrox*. *Journal of Herpetology* 17, 265 (1983).

98 Tolson, P. J., The reproductive biology of the neotropical boid genus *Epicrates* (Serpent: Boidae), in *Reproductive Biology of South American Vertebrates*, edited by W. C. Hamlett (Springer-Verlag, New York, 1992), p. 165.

99   Gillingham, J. C., Social behavior, in *Snakes: Ecology and Evolutionary Biology*, edited by R. A. Siegel, J. T. Collins, and S. S. Novak (McGraw-Hill, New York, 1987), p. 184.

100 Schwartz, J. M., McCracken, G. F., and Burghardt, G. M., Multiple paternity in wild population of the garter snake *Thamnophis sirtalis*. *Behavioral Ecology and Sociobiology* 25, 269 (1989).

101 Weatherhead, P. J. and Robertson, R. J., Offspring quality and the polygyny threshold: the sexy son hypothesis. *The American Naturalist* 138, 1159 (1979).

102 Halliday, T. and Arnold, S., Multiple mating by females: a perspective from quantitative genetics. *Animal Behaviour* 35, 939 (1987).

103 Eberhard, W. G., *Female Control: Sexual Selection by Cryptic Female Choice* (Princeton University Press, Princeton, NJ, 1996).

104 Madsen, T. and Shine, R. G., Male mating success and body size in European grass snake. *Copeia* 1993, 561 (1993).

105 Weatherhead, P. J., Barry, F. E., Brown, G. P., and Forbes, M. R. L., Sex ratios, mating behavior and sexual size dimorphism of northern water snake, *Nerodia sipedon*. *Behavioral Ecology and Sociobiology* 36, 301 (1995).

106 King, R. B., Bitner, T. D., Queral, R. A., and Cline, J. H., Sexual dimorphism in neonate and adult snakes. *Journal of Zoology* 247, 1928 (1999).

107 Madsen, T. and Shine, R. G., Costs of reproduction influence the evolution of sexual size dimorphism in snakes. *Evolution* 48, 1389 (1994).

108 Luiselli, L., Individual success in mating balls of the grass snake, *Natrix natrix*: size is important. *Journal of Zoology* 239, 731 (1996).

109 Madsen, T. and Shine, R. G., Quantity or quality? Determinants of maternal reproductive success in tropical pythons (*Liasis fuscus*). *Proceedings of the Royal Society of London Series B Biological Sciences* 265, 1521 (1998).

110 Shine, R. G. et al., Chemosensory cues allow courting male garter snakes to assess body length and body condition of potential mates. *Behavioral Ecology and Sociobiology* 54, 162 (2003).

111 Rivas, J. A., Feasibility and efficiency of transmitter force-feeding in studying the reproductive biology of large snakes. *Herpetological Natural History* 2001, 8(1):93-95.

112 Arnqvist, G., On multiple mating and female fitness: comments on Loman et al. (1988). *Oikos* 54, 248 (1989).

113 Stille, B. T., Madsen, T., and Niklasson, M., Multiple paternity in the adder, *Vipera berus*. *Oikos* 47, 173 (1986); Höggren, M. and Tegelström, H., DNA fingerprinting shows within-season multiple paternity in the adder (*Vipera berus*). *Copeia* 1995, 271 (1995); McCracken, G. F., Burghardt, G. M., and Houts, S. E., Microsatellite markers and multiple paternity in the garter snake *Thamnophis sirtalis*. *Molecular Ecology* 8, 1475 (1999).

114 Barry, F. E., Weatherhead, P. J., and Phillip, D. P., Multiple paternity in a wild population of northern water snake, *Nerodia sipedon*. *Behavioral Ecology and Sociobiology* 30, 193 (1992).

115 Devine, M. C., Copulatory plugs in snakes: enforced chastity. *Science* 187, 844 (1975); Devine, M. C., Potential for sperm competition in reptiles: behavioral and physiological consequences, in *Sperm Competition and the Evolution of Animal Mating Systems*, edited by R. L. Smith (Academic Press, Orlando, FL, 1984), p. 509.

116 Schuett, G. W. and Gillingham, J. C., Courtship and mating of the copperhead, *Agkistrodon contortix*. *Copeia* 1988, 374 (1988).

117   Perry-Richardson, J. J., Schofield, C. W., and Ford, N., Courtship of garter snake, *Thamnophis marcianus*, with a description of female behavior for coitus interruption. *Journal of Herpetology* 24, 76 (1990).

118   Shine, R. G., Langkilde, T., and Mason, R. T., Cryptic forcible insemination: male snakes exploit female physiology, anatomy and behavior to obtain coercive matings. *The American Naturalist* 162 (5), 653 (2003); Shine, R. G., Langkilde, T., and Mason, R. T., Battle of sexes: forcibly-inseminating male garter snakes target courtship to more vulnerable females. *Animal Behaviour* 70, 1133 (2005); Shine, R. G. and Mason, R. T., Does large body size in males evolve to facilitate forced insemination? A study on garter snakes. *Evolution* 59, 2426 (2005).

119   Wesneat, D. F., Reproductive physiology and sperm competition in birds. *Trends in Ecology and Evolution* 11, 191 (1996).

120   Zahavi, A., Mate selection for a handicap. *Journal of Theoretical Biology* 53, 205 (1975).

121   Duvall, D. G., Arnold, S., and Schuett, G. W., Pitviper mating system: ecological potential, sexual selection, and microevolution, in *Biology of Pitvipers*, edited by J. A. Campbell and E. D. Brodie (La Selva, Tyler, TX, 1992), p. 321.

122   Shine, R. G. and Fitzgerald, M., Variation in mating systems and sexual size dimorphism between populations of Australian python *Morelia spilota* (Serpentes: Pythonidae). *Oecologia* 103, 490 (1995).

123   Frieddemann, N. S. de and Arocha, J., Herederos del Jaguar y la Anaconda, in *Herederos del Jaguar y la Anaconda* (Biblioteca Nactional de Colombia, Bogotá, 2016).

124   Hoorn, C., Guerrero, J., and Lorente, M. A., Andean tectonics as a cause for changing drainage patterns in Miocene northern South America. *Geology* 23 (3), 237 (1995); Lundberg, J. G. et al., The stage for neotropical fish diversification: a history of tropical South American rivers, in *Phylogeny and Classification of Neotropical Fishes*, edited by L. R. Malabarba et al. (Edipucrs, Porto Alegre, 1998), p. 13; Latrubesse, E. M. et al., Late Miocene paleogeography of the Amazon Basin and the evolution of the Amazon River system. *Earth Science Reviews* 99, 99 (2010).

125   Díaz de Gamero, M., The changing course of the Orinoco River during the Neogene: a review. *Palaeogeography, Palaeoclimatology, Palaeoecology* 123, 385 (1996).

126   Wesselingh, F. P. and Hoorn, C., Geological development of Amazon and Orinoco basins, in *Historical Biogeography of Neotropical Freshwater Fishes*, edited by J. S. Albert and R. E. Reis (University of California Press, Los Angeles, 2011), p. 59.

127   Mora, A. et al., Tectonic history of the Andes and sub-Andean zones: implications for the development of the Amazon drainage basin, in *Amazonia, Landscape and Species Evolution: A Look into the Past*, edited by C. Hoorn and F. P. Wesselingh (Blackwell Publishing, Oxford, 2010), p. 38.

128   Mittermeier, R. et al., The Pantanal, in *Wilderness*, edited by R. Mittermeier (Cemex, Mexico, 2002).

129   Strecker, M. R. et al., Tectonic and climate of the Southern Central Andes. *Annual Review of Earth and Planetary Science* 35, 747 (2006); Vonhof, H. B. and Kaandorp, R. J. G., Climate variation in the Amazonia during neogene and the quaternary, in *Amazonia: Landscape and Species Evolution*, edited by C. Hoorn and F. Wesselingh (Wiley-Blackwell, Oxford, 2010), p. 201.

130   Mörner, N.-A., Origin of the Amazonian rainforest. *International Journal of Geosciences* 7, 470 (2016).

131   Hoorn, C., Wesselingh, F., Hovikoski, J., and Guerrero, J., Development of the Amazonian mega-wetland (Miocence; Brazil, Colombia, Peru, Bolivia), in *Amazonia: Landscape and Species Evolution*, edited by C. Hoorn and F. Wesselingh (Wiley-Blackwell, Oxford, 2010), p. 123.

132   Dirksen, L. and Böhme, W., Studies on anacondas III: a reappraisal of *Eunectes beniensis* Dirksen, 2002, from Bolivia, and a key to the species of the genus *Eunectes* Wagler, 1830 (Serpentes: Boidae). *Russian Journal of Herpetology* 12, 223 (2005).

133   Noonan, B. P. and Chippindale, P. T., Dispersal and vicariance: The complex evolutionary history of boid snakes. *Molecular Phylogenetics and Evolution* 40, 347 (2006); Colston, T. J. et al., Molecular systematics and historical biogeography of tree boas (*Corallus* spp.). *Molecular Phylogenetics and Evolution* 66 (3), 953 (2013).

134   Hsiou, A. S. and Albino, A. M., Presence of the Genus Eunectes (Serpentes, Boidae) in the Neogene of Southwestern Amazonia, Brazil. *Journal of Herpetology* 43 (4), 612 (2009); Albino, A. M., The South American fossil squamatta (Reptilia: Lepidosauria), in *Contributions of Southern South America to Vertebrate Paleontology*, edited by G. Arratia (Verlag, Munchen, 1996), Vol. 30; Dr. Friedrich Pfeil, p. 185.

135   King Jr., Martin Luther, *Beyond Vietnam—A Time to Break Silence* (https:// americanrhetoric.com/speeches/mlkatimetobreaksilence.htm, 1967).

136   McSweeney, K., Indigenous population growth in the lowland Neotropics: social science insights from biodiversity conservation. *Conservation Biology* 19 (5), 1375 (2005).

137   Vickers, W. T., Hunting yields and game composition over ten years in an Amazon Indian territory, in *Neotropical Wildlife Use and Conservation*, edited by J. G. Robinson and K. H. Redford (University of Chicago Press, Chicago, 1991), p. 53; Fitzgerald, L. A., Chani, J. M., and Donadio, O. E., Tupinambis lizards in Argentina: implementing management of a traditionally exploited resource, in *Neotropical Wildlife Use and Conservation*, edited by J. G. Robinson and K. H. Redford (Chicago University Press, Chicago, 1991), p. 303; Robinson, J. G. and Redford, K. H., *Neotropical Wildlife Use and Conservation* (University of Chicago Press., Chicago, 1991).

138   Camhi, M., Industrial fisheries threaten ecological integrity of the Galapagos Islands. *Conservation Biology* 9 (4), 715 (1995).

139   Rivas, J. A., Is wildlife management business or conservation—a question of ideology. *Reptiles and Amphibians* 17 (2), 112 (2010).

140   Ojasti, J., Human exploitation of capybaras, in *Tropical Wildlife Use and Conservation*, edited by J. G. Robinson and K. H. Redford (University of Chicago Press, Chicago, 1991), p. 236.

141   Waller, T., Micucci, P. A., Mendez, M., and Alvarenga, E., The management and conservation biology of yellow anacondas (*Eunectes notaeus*) in Argentina, in *Biology of Boas, Pythons, and Related Taxa*, edited by R. W. Henderson, R. Powell, G. W. Schuett, and M. E. Douglas (Eagle Mountain Publishing, Eagle Mountain, UT, 2007).

142   Thorbjarnarson, J. B., Messel, H., King, F. W., and Ross, J. P., *Crocodiles: An Action Plan for Their Conservation* (IUCN, Gland, Switzerland, 1992).

143   Holmstrom, W. F., *Eunectes* (anacondas): maturation. *Herpetological Review* 13, 126 (1982).

144   Hoover, C., *The US Role in the International Live Reptiles Trade: Amazon Tree Boas to Zululand Dwarf Chameleons* (TRAFFIC North America, Washington, DC; https://www.traffic.org/site/assets/files/9703/the-us-role-in-the-international-live-reptile-trade-1.pdf, 1998). Dorcas, M. E. et al., Severe mammal declines coincide with proliferation of invasive Burmese pythons in Everglades National Park. *Proceedings of the National Academy of Sciences* **109** (7), 2418 (2012).

145   Atkinson, I., Introduced animals and extinctions, in *Conservation Biology for the Twenty-First Century*, edited by D. Western and M. Pearl (Oxford University Press, New York, 1989).

146   Thorbjarnarson, J. B., Crocodile tears and skins: international trade, economic constraints and limits to sustainable use of crocodilians. *Conservation Biology* **13**, 465 (1999).

147   Caughley, G., *Analysis of Vertebrate Populations* (John Wiley and Sons, New York, 1977); Caughley, G. and Sinclair, A. R., *Wildlife Ecology and Management* (Blackwell Scientific Publications, Oxford, 1994).

148   Thorbjarnarson, J. B. and Velasco, A., Economic incentives for management of Venezuelan caiman. *Conservation Biology* **13**, 397 (1999).

149   Shine, R. G., Ambariyanto, Harlow, P. S., and Mumpuni, Reticulated python in Sumatra: biology, harvesting and sustainability. *Biological Conservation* **87**, 349 (1999).

150   Joanen, T., McNease, L., Elsey, R. M., and Staton, M. A., The commercial consumptive use of the American alligator (*Alligator mississippiensis*) in Louisiana: its effect in conservation, in *Harvesting Wild Species: Implications for Biodiversity Conservation*, edited by C. H. Freese (Johns Hopkins University Press, Baltimore, MD, 1997), p. 465; Medellín, R. A., Sustainable harvest for conservation. *Conservation Biology* **13**, 225 (1999); McLarney, W. O., Sustainable development: a necessary means for the effective biological conservation. *Conservation Biology* **13**, 4 (1999); Struhsaker, T. T., A biologist's perspective in the role of sustainable harvest in conservation. *Conservation Biology* **12**, 930 (1998); Robinson, J. G., The limits of caring: sustainable living and the loss if biodiversity. *Conservation Biology* **7**, 20 (1993).

151   Rodríguez, J. P. and Rojas-Suarez, F., Guidelines for the design of conservation strategies for the animals of Venezuela. *Conservation Biology* **10**, 1245 (1996).

152   Rivas, J. A., The miracle of universities. *Conservation Biology* **12** (6), 1169 (1998).

153   Rivas, J. A. and Owens, R. Y., Teaching conservation effectively: a lesson from life history strategies. *Conservation Biology* **13** (2), 453 (1999).

154   Whitten, T., Holmes, D., and Mackinnon, K., Conservation biology: a displacement behavior for academia? *Conservation Biology* **15** (1), 1 (2001).

155   Mansfield, B., Sustainability, in *A Companion to Environmental Geography*, edited by N. Castree, D. Demeritt, D. Liverman, and B. Rhoads (Blackwell Publishing Ltd., 2009), p. 37.

156   Thorbjarnarson, J., Analysis of the spectacled caiman (*Caiman crocodilus*) harvest program in Venezuela, in *Neotropical Wildlife Use and Conservation*, edited by J. G. Robinson and K. H. Redford (Chicago University Press, Chicago, 1991), p. 217.

157   Kütting, G., *Globalization and the Environment* (State University of New York Press, Albany, 2004).

158   World Bank. *Adjustment lending retrospective: final report operation policies and country services.* (World Bank, Washington, DC, 2001).

159 Clapp, J. and Dauvergne, P., Path to a green world, in *Path to a Green World* (MIT Press, Cambridge, MA, 2005).

160 Navarro-Jimenez, G., *Geopolítica imperialista. De la doctrina de los dos hemisferios a la doctrina imperial de George W. Bush* (Ediciones Zitra, Quito, 2004).

161 Navarro-Jimenez, G., La democracia y el "fin de la historia." *Red Voltaire* (Voltairenet. org, 2005); Jochnick, C., Perilous prosperity. *New Internationalist* **335** (http://www. thirdworldtraveler.com/South_America/Perilous_Prosper_Ecuador.html, 2001).

162 Buhdoo, D., IMF/World Bank wreak havoc on the Third World, in *50 Years Is Enough: The Case Against the World Bank and the International Monetary Fund*, edited by K. Danaher (South End Press, Boston, 1994), p. 20.

163 Rich, B., *Mortgaging the Earth: The World Bank, Environmental Impoverishment and the Crisis of Development* (Earthscan, London, 1994).

164 Goldzimer, A., Worse than the World Bank? Export credit agencies, the secret engine of globalization. *Food First Backgrounder* **9** (1) (2003).

165 Perkins, J., *Confessions of an Economic Hit Man* (Berret-Koehler Publisher Inc., San Francisco, 2004) p. 284.

166 Horta, K., Multilateral development institutions bear major responsibility on solving Africa's environmental problems, in *Environmental Policies for Sustainable Growth in Africa. Proceedings from the Fifth International Conference*, edited by P. LeBel (State University of Upper Montclair, NJ, 1991), p. 21.

167 Yunus, M., Preface: redefining development, in *50 Years Is Enough: The Case Against the World Bank and the International Monetary Fund*, edited by K. Danaher (South End Press, Boston, 1994), p. ix.

168 Cheru, F., 1992 structural adjustment, primary source trade and sustainable development in sub-Saharan Africa. *World Development* **20** (4), 497 (1992).

169 Blustein, P., *And the Money Kept Rolling In (and Out): Wall Street, the IMF and the Bankrupting of Argentina* (Public Affairs, New York, 2005).

170 Clapp, J., Toxic exports, the transfer of hazardous wastes from rich to poor countries, in *Toxic Exports, the Transfer of Hazardous Wastes from Rich to Poor Countries* (Cornell University Press, Ithaca, NY, 2001) .

171 Pearce, D., Adger, N., Maddison, D., and Moran, D., Debt and the environment. *Scientific American* **272** (June), 52 (1995); Horta, K., Round, R., and Young, Z., *The Global Environmental Facilities: The First Ten Years—Growing Pains or Inherent Flaws?* (Environmenta), 36.

172 Forero, J., Seeking balance: growth vs culture in the Amazon. *New York Times*, 2003; Ellin, A., Suit says Chevron Texaco Dumped poisons in Ecuador. *New York Times*, 2003.

173 Rogoff, K., This time is not different. *Newsweek (Atlantic Ed)* (February 2004).

174 Danaher, K. ed., *50 Years Is Enough: The Case Against the World Bank and the International Monetary Fund* (South End Press, Boston, 1994).

175 Larrea, M., *El referendo y la descolonización de Venezuela* (http://www.voltairenet. org/article122533.html?var_recherche=Caracazo?var_recherche=Caracazo, 2004).

176 World Bank, *World Development Report* (Oxford University Press, New York, 1992).

177 Singer, P. W., *Corporate Warriors: The Rise of the Privatized Military Industry* (Cornell University Press, Ithaca, NY, 2003).

178 Gollinger, E., *The Chávez Code* (Public Affairs, New York, 2005); Lemoine, M., Haiti: Titide's downfall. *Le Monde diplomatique* (http://mondediplo.com/2004/09/09haiti?var_recherche=Jean+Bertrand+Aristide, 2004).

179 McAfee, K., Jamaica: the showpiece that didn't stand up. in *50 Years Is Enough: The Case Against the World Bank and the International Monetary Fund*, edited by K. Danaher (South End Press, Boston, 1994), p. 68.

180 Czech, B., Economic growth as the limiting factor for wildlife conservation. *Wildlife Society Bulletin* 28, 4 (2000); Czech, B., Krausman, P. R., and Devers, P. K., Economic associations among causes of species endangerment in the United States. *BioScience* 50, 593 (2000).

181 Friedlingstein, P. et al., Update on CO_2 emissions. *Nature Geoscience* 3, 811 (2010).

182 Oda, T., and Maksyutov, S., A very high-resolution (1km×1 km) global fossil fuel CO_2 emission inventory derived using a point source database and satellite observations of nighttime lights. *Atmospheric Chemistry and Physics* 2011, 11:543-556.

183 Boden, T. A., Marland, G., and Andres, R. J., Global, Regional, and National Fossil-Fuel CO_2 Emissions In.: Carbon Dioxide Information Analysis Center, Oak Ridge National Laboratory, U.S. Department of Energy Oak Ridge, Tenn., U.S.A.; 2013.

184 Donlan, C. J. et al., Pleistocene rewilding: an optimistic agenda for twenty-first century conservation. *The American Naturalist* 168 (5), 660 (2006).

185 Kareiva, P. and Marvier, M., What is conservation science? *BioScience* 62 (11), 962 (2012).

186 Miller, B., Soulé, M. E., and Terborgh, J., "New conservation" of surrender to development. *Animal Conservation* 17 (4), 509 (2014).

187 Rivas, J. A., Environmental conservation and socialism. a conservationist manifesto for the Venezuela's revolution. *Axis of Logic* (http://www.axisoflogic.com/artman/publish/Article_25003.shtml, 2007).

188 Aimard, G., *The Freebooters: A Story of the Texan War* (Otbebookpublishing, Chicago, 2017).

189 Laclau, E. and Mouffe, C., *Hegemony and Socialist Strategy: Towards a Radical Democratic Politics*, 2nd ed. (Verso, New York, 2001).

190 Berry, W., Thoughts in the presence of fear. *Orion* (http://www.orionsociety.org/pages/oo/sidebars/America/Berry.html, 2001).

191 Machado, A., *Poemas completos*, 11th ed. (Espasa-Calpe, Madrid, 1936).

192 Rivas, J. A., Dieta de la iguana verde (*Iguana iguana*) en los llanos centrales de Venezuela durante la estación seca. Undergraduate thesis presented to Universidad Central de Venezuela, 1990.

193 Rivas, J. A. and Levin, L., Sex differential antipredator behavior in juvenile green iguanas, *Iguana iguana*: evidences for fraternal care, in *Behavior, Diversity, and Conservation of Iguanas*, edited by A. C. Alberts, R. L. Carter, W. K. Hayes, and E. P. Martins (University of California Press, Berkeley, 2002), p. 119.

194 Rivas, J. A. and Ávila, T. M., Sex differences in hatchling *Iguana iguana*. *Copeia* 1996, 219 (1996).

195 Burghardt, G. M., Human–bear bonding in research on black bear behavior, in *The Inevitable Bond: Examining Scientist–Animal Interactions*, edited by H. Davis and D. Balfour (Cambridge University Press, Cambridge, 1992), 365.

Index

Figures and tables are indicated by *f* and *t* following page numbers. The color plates are indicated by number following the word Plate(s).